SENSING
SOUND

Sign, Storage, Transmission • *A series edited by Jonathan Sterne and Lisa Gitelman*

NINA SUN EIDSHEIM

SENSING
SOUND

Singing & Listening as Vibrational Practice

Duke University Press · Durham and London · 2015

© 2015 NINA SUN EIDSHEIM. All rights reserved

Designed by Courtney Leigh Baker
Typeset in Whitman and Gill Sans by Tseng Information Systems, Inc.

Library of Congress Cataloging-in-Publication Data
Eidsheim, Nina Sun, [date] author.
Sensing sound : singing and listening as vibrational practice / Nina Sun Eidsheim.
pages cm — (Sign, storage, transmission)
Includes bibliographical references and index.
ISBN 978-0-8223-6046-9 (hardcover)
ISBN 978-0-8223-6061-2 (pbk.)
ISBN 978-0-8223-7469-5 (e-book)
1. Sound. 2. Singing. 3. Vibration. 4. Music—Acoustics and physics.
I. Title. II. Series: Sign, storage, transmission.
ML3807.E43 2015
781.1—dc23   2015022741

COVER ART: Vilde Rolfsen, *Plastic Bag Landscape*. Courtesy of the artist.

Duke University Press gratefully acknowledges the support of the AMS 75 PAYS Endowment of the American Musicological Society, funded in part by the National Endowment for the Humanities and the Andrew W. Mellon Foundation, which provided funds toward the publication of this book.

IN MEMORY OF & DEDICATED TO
Hillary Elizabeth Brown (1971–2011) · Nicolás Arnvid Henao Eidsheim (2011–)

CONTENTS

Illustrations · viii
Acknowledgments · xi
Introduction · 1

1 **MUSIC'S MATERIAL DEPENDENCY**
What Underwater Opera Can Tell Us about Odysseus's Ears · 27

2 **THE ACOUSTIC MEDIATION OF VOICE, SELF, AND OTHERS** · 58

3 **MUSIC AS ACTION**
Singing Happens before Sound · 95

4 **ALL VOICE, ALL EARS**
From the Figure of Sound to the Practice of Music · 132

5 **MUSIC AS A VIBRATIONAL PRACTICE**
Singing and Listening as Everything and Nothing · 154

Notes · 187
Bibliography · 241
Index · 261

1.1	Juliana Snapper singing underwater · 28
1.2	Ron Athey on the Judas cradle · 30
1.3	Juliana Snapper singing upside down in *Judas Cradle* · 38
1.4	Juliana Snapper singing in bathtub · 42
1.5	Snapper singing in water tank · 42
1.6	Snapper with two tenders · 43
1.7	Eidsheim and Bieletto in pool · 44
2.1	Audible and acoustic factors · 67
2.2	*Songs of Ascension*, Oliver Ranch, Geyserville, CA · 73
2.3	*Songs of Ascension*, Stanford University, Palo Alto, CA · 76
2.4	*Songs of Ascension*, Guggenheim Museum, New York, NY · 76
2.5	*Songs of Ascension*, Disney Hall, Los Angeles, CA · 77
2.6	Map of Union Station, Los Angeles, CA · 83

ILLUSTRATIONS

2.7   Overture of *Invisible Cities*, Union Station · 84
2.8   Dancers during performance of *Invisible Cities* · 86
2.9   *Invisible Cities*, rehearsal · 86
2.10 Singer with cellphone; audience with headset · 88
3.1   SpeechJammer · 98
3.2   Three *Noisy Clothes* costumes · 106
3.3   Person bending down; person standing · 107
3.4   Silhouettes of clothes · 107
3.5   Early list of body movements, *Body Music* · 114
3.6   Early abandoned sketch, *Body Music* · 117
3.7   Draft of section of final iteration of *Body Music* · 119
5.1   Wheel of Acoustics · 166
5.2   Vibratory Model of the Human Body · 173

ACKNOWLEDGMENTS

The process of conceiving and writing this book is a testament to its thesis that sound does not exist in a vacuum, but rather comes into existence through particular and always already unique material iterations. In the same way, any ideas expressed herein came about within a communal environment — whether through interactions with scholarly discourses and citational frameworks or through conferences, talks, and personal communications. Moreover, as I finally face the task of writing the acknowledgments I realize that, like the rich phenomenon of music, the gratitude I feel toward all the individuals and institutions that supported me throughout this process cannot adequately be captured in words. However, for their tremendous support and enormously helpful suggestions, I do want to mention some individuals by name. Needless to say, the idiosyncrasies that remain are mine.

First, many thanks to my editor, Ken Wissoker, for truly understanding and trusting in this project. Thanks also to Jade Brooks and Danielle Szulczewski for expertly bringing the manuscript through the process; and to Jeanne Ferris for wonderful copy editing. And to Jonathan Sterne and Joseph Auner for their tremendous work in reviewing the manuscript and for revealing their identities to me to enable and expand the conversation.

Special thanks to my colleagues in the Department of Musicology at the University of California, Los Angeles (UCLA): Olivia Bloechl, Robert Fink, Raymond Knapp, Elisabeth Le Guin, Tamara Levitz, David MacFayden, Mitchell Morris, Jessica Schwartz, Timothy Taylor, and Elizabeth Upton; and to graduate students at UCLA and beyond (especially Alexandra Apolloni, Robbie Beahrs, Natalia Bieletto, Ben Court, Oded Erez, Hyun Kyong Chang, Rebecca Lippman, Joanna Love, Caitlin Marshall, Andrea Moore, Tiffany Naiman, David Utzinger, and Schuyler Whelden; and to Breena Loraine, Mike D'Errico, Jillian Rogers, Zachary Wallmark, and Mandy-Suzanne Wong for working closely with me on multiple projects. Thanks are also due to the exceptional two mentors assigned to me by the UCLA Council of Advisors, Joseph Bristow and Anastasia Loukaitous-Sideris; to Joy Doan, David Gilbert, and David Gilbert at the UCLA Music Library; to Barbara van Nostrand, Olivia Diaz, and the rest of the humanities administrative group; the UCLA Herb Alpert School of Music staff; and Assistant Dean of Humanities Reem Hanna-Harwell and Director of Academic Personnel and Operations Lauren Na at UCLA who together make everything possible.

Colleagues I have spent loads of time with, cooking up and carrying out large projects in the service of forwarding the conversation and possibilities for expanding research discourse around voice, include Annette Schlichter, in our collaborations convening research groups (the UC Multicampus Research Group [MRG] titled Keys to Voice Studies: Terminology, Methodology, and Questions across Disciplines and the UC Humanities Research Center Residency Research Group entitled Vocal Matters: Technologies of Self and the Materiality of Voice) and co-editing the forthcoming special issue of *Postmodern Culture* on voice and materiality; Jody Kreiman, Zhaoyan Zhang, Rosario Signorello, and Bruce Garrett for being willing to answer endless questions about voice and vibration and for imagining what voice studies could one day be at UCLA; and Katherine Meizel for taking on the significant editorial and organizational work of *The Oxford Handbook of Voice Studies* and its related conference, "Voice Studies Now," with me.

For generously engaging me in conversation and sharing resources at critical junctures, I thank Shane Butler, Paul Chaikin, J. Martin Daughtry, Joanna Demers, Emma Dillon, Ryan Dohoney, Emily Dolan, Veit Erlman, David Gutkin, Juliana Hodkinson, David Howes, Brandon LaBelle, Douglas Kahn, Brian Kane, Alejandro Madrid, Susan McClary, Mara Mills, Matthew Morrison, Jamie Niesbet, Marina Peterson, Benjamin Piekut, Matthew Rahaim, Juliana Snapper, Jason Stanyek, Alexander Weheliye, Amanda Weidman, Rachel Beckles Willson, and Maite Zubiaurre.

To Daphne Brooks for inviting me to be part of the Black Feminist Sonic Studies Group, and to its stellar lineup of Farah Jasmine Griffin, Emily Lordi, Mendi Obadike, Imani Perry, Salamishah Tillet, and Gayle Wald; to members of the UC MRG (especially Theresa Allison, Christine Bacareza Balance, Robbie Beahrs, Shane Butler, Julene Johnson, Patricia Keating, Sarah Kessler, Peter Krapp, Jody Kreiman, Caitlin Marshall, Miller Puckette, Annelie Rugg, Mary Ann Smart, James Steintrager, and Carole-Anne Tyler); to the UC Humanities

Research Center Residency Research Group (Jonathan Alexander, David Kasunic, Katherine Kinney, Caitlin Marshall, and Carole-Anne Tyler); to the Cornell University Society for the Humanities (Eliot Bates, Marcus Boon, Duane Corpis, Miloje Despic, Sarah Ensor, Ziad Fahmy, Brian Hanrahan, Michael Jonik, Jeannette S. Jouili, Damien Keane, Nicholás Knouf, Brandon LaBelle, Eric Lott, Roger Moseley, Norie Neumark, James Nisbet, Trevor Pinch, Jonathan Skinner, Jennifer Stoever-Ackerman, and Emily Thompson); and to participants invited to the "Vocal Matters: Embodied Subjectivities and the Materiality of Voice" symposium (Joseph Auner, Charles Hirschkind, Mara Mills, Jason Stanyek, Jonathan Sterne, and Alexander Weheliye)—thank you!

Many of the ideas herein were first presented in talks and roundtables. I thank all of those who have engaged me in questions and conversation. For invitations to speak about voice and vibration, I thank Ryan Doheney and Hans Thomalla and the Northwestern University School of Music; Paul Sommerfeld at Duke University and the members of the South Central Graduate Music Consortium; Stan Hawkins and the University of Oslo; Zeynep Bulut and the Institute for Critical Inquiry Berlin; Daphne Brooks and the Princeton Center for African American Studies; Dylan Robinson, Sherrie Lee, and the University of Toronto; Robbie Beahrs and Benjamin Brinner at the UC Berkeley Department of Music; Martha Feldman and David Levin at the University of Chicago Neubauer Collegium for Culture and Society; Catherine Provenzano and J. Martin Daughtry at the New York University; Jann Pasler and the UC San Diego Department of Music; Konstantinos Thomaidis and Ben Macpherson at the Centre for Interdisciplinary Voice Studies; the Society for Ethnomusicology; the American Musicological Society; and the International Conference Crossroads in Cultural Studies.

While this project did not originate with my dissertation, which treated issues related to vocal timbre and race, I would be remiss if I did not recognize the intellectual influence of key people from my graduate student years and on: Jann Pasler, George Lewis, John Shepherd, Miller Puckette, Adriene Jenik, George Lipsitz, Deborah Wong, Andy Fry, Steven Schick, Juliana Hodkinson, Jacqueline and Mark Bobak, Paul Berkolds, and the late Ernest Fleischmann and James Tenney. And, much earlier, the influence of Gayle Opaas, Tor Strand, Atle Færøy, and Anne-Brit Krag.

I experience a special kind of gratitude for the amazing writing communities of which I am part. For sustenance, sanguine advice, and good laughs, my thanks go to: Sara, Muriel, Katherine, Leslie, Juliana, Lauri, Jessica, Carrie, Julie, Ray, Sherie, David, Tracy, Kathy, Emily, Tavishi, and Jørgen. Similarly, to my spirited collaborators Elodie Blanchard, Pai Chou, Luis Fernando Henao, Alba Fernanda Triana, and Sandro del Rosario. And to Tildy Bayar, Mandy-Suzanne Wong, Jane Katz, Shane Butler, and Sara Melzer for intense reading and commenting on part or all of this manuscript; and especially to William Waters for reading the entire manuscript multiple times at different stages of completion.

For the patchwork of contemporary family village life that we have managed to stitch together in the United States, I am forever grateful to *onkel* Phillip; Lolly and Gary; Olivia and Sophia; Selene and Lauren; April, Bob, and Lucas; Julie, Tony, and Seth; Rosa in Los Angeles; Lindsay and family in San Francisco; Erle and Peggy in Arlington; Alba and Jose in Miami; and Alexandra and family in New York. To our incredible family in Colombia: Alba Lucia, Karina, Luis Darienze, and Laurita; Adriana, Enrique, and Camila; Mariluz, Luna, and Lukas; and, especially to *mi suegras*, Amparo and Gustavo, *por toda su paciencia y gran ayuda ya que este libro fue en progreso. Muchas gracias por todo*. And to our equally patient and supportive family and friends in Norway: Marianne *med familie*; Jørgen; *tante* Aashild *og mostemann* Arve; Sam, Ingrid, Aurora, Sunniva, Lill Beate, mamma *og* pappa, *og* mormor. *Tusen millioner takk!*

To Nicolás for teaching me uncountable new vocal moves and a thing or two about intermaterial vibrations; and, finally, to Luisfer—whose practice of patience, kindness, and love carries our family through every day.

A MUCH EARLIER FORM of parts of chapter 1 has appeared elsewhere: in "Sensing Voice: Materiality and the Lived Body in Singing and Listening" in *Senses & Society* 6, no. 2 (2011), with permission from Bloomsbury Publishing Plc., and in *Voice Studies: Critical Approaches to Process, Performance and Experience*, Konstantinos Thomaidis and Ben Macpherson, editors (New York: Routledge, 2015).

For permission to reproduce images, I thank Marina Ancona, Elodie Blanchard, Miha Fras, Stephanie Berger/The New York Times/Redux, Axel Koester, Kazutaka Kurihara and Koji Tsukada, Maria Mikheyenko, Jill Rogers, Dana Ross, Yuval Sharon, Silvana Torrinha, and Alba Triana.

My research was supported by a UCLA Council of Research Grant, a UC Institute for Research in the Arts Performance Practice and Arts Grant, a UCLA Research Enabling Grant, the Miles Levin Essay award at the Mannes Institute on Musical Aesthetics, and a UCLA Center for the Study of Women Faculty Research Grant. In addition, I received support from the Woodrow Wilson

Mellon Foundation; the Cornell University Society for the Humanities; the Department of Musicology at UCLA; the Office of the Dean of Humanities at UCLA; and the AMS 75 PAYS Endowment of the American Musicological Society, funded in part by the National Endowment for the Humanities and the Andrew W. Mellon Foundation.

INTRODUCTION

You may not remember the first time you heard the query, or how many times you have heard it since: "If a tree falls in the forest and no one is there to hear it, does it make a sound?" Usually, people pose this conundrum to raise questions about reality and observation.[1] However, having mulled it over for quite some time, I think that the question's import lies elsewhere. If you were there in the forest, the sound of the falling tree might be one of your lesser concerns. Your attention might be drawn to the darkening of the sky as the great tree crashes down, filling your visual horizon. You might notice the eerie sounds of birds as they flee; perhaps you would squint as your eyes burned from the dust that whirled upward, saturating the air; or you might feel alarmed by the thump of the tree crashing to the ground through the branches of other trees, even bringing them down with it. You might simply be overwhelmed by the impact of the thump vibrating through your body. Conceiving of a falling tree as sound alone does not even begin to address the phenomena that are involved. The same applies to music, sound, singing, and listening.

For Clifford Geertz, an ethnographic scene deserves a "thick description" so that we can begin to tease out its intent and the meaning involved. Writing about an event so apparently unambiguous as the flick of an eye, Geertz distinguished between a wink, a twitch, and the imitation of a wink.[2] Analogously, just as an ethnographic interpretation might fail to take account of the local culture and context within which the event is taking place, interpreting a sense experience in terms of just one of the physical senses cannot take full account of the event's complexities.

The fact that the "thick" event of the falling tree elicits a question about sound may be instructive in multiple ways, speaking not only to issues in music discourse and scholarship but also to a broader tendency regarding complex sensory phenomena. The question concerning the tree, and the kinds of questions we ask concerning music, are symptomatic of a propensity to reduce thick events to manageable signifiers. On the one hand, this could be understood simply as a general cognitive strategy that enables us to deal with and move through a complex world. On the other hand, it is nevertheless important to be constantly aware of the ways in which shifting forces and dynamics of power inscribe themselves onto the perspectives and processes of this reduction.

Sonic reductions—that is, the tendency to constrain our understanding of sound through previously defined referents—arise from assumptions and values concerning the usefulness of sound in constructing meaning.[3] That is, we rely on the phenomena that we broadly conceptualize as *sound* to be stable, carrying out the work we need them to accomplish—for example, in something as commonplace as distinguishing between sound and noise, or sound and music, or noise and music. (In chapters 2, 3, and 4, I discuss in more detail the kinds of work that we rely on sound to carry out.) Certainty regarding a given sound and its meaning relies on the premise that a thick sonic event may be reduced to a static one, and in the process of this reduction we identify an object, a stable referent. As a result, the thick event of music is understood through restricted and fixed notions such as pitch, durational schemes, forms, genres, and so on—and thus the dynamic, multifaceted, and multisensorial phenomenon of sound is often reduced to something static, inflexible, limited, and monodimensional. Music, then, is most commonly experienced through tropes, or what I call the *figure of sound*.[4] With this term I attempt to capture the process of ossification, through which I argue that an ever-shifting, relationally dependent phenomenon comes to be perceived as a static object or incident. It is precisely because the figure of sound is, by definition, a naturalized concept that inquiries into voice and music, which are based on it, are similarly defined.

Through reconceptualizing the voice as an object of knowledge—and, relatedly, through investigating voice and music as intermaterial practices—we may begin to understand that voice and the states it has to offer are multifaceted and sometimes contradictory. Thus, I suggest that through the insights gleaned from taking the voice seriously as an object of knowledge, we may release music and sound from its containment within a limited set of senses and fixed meanings. Hence, music's ontological status can be changed from an external, knowable object to an unfolding phenomenon that arises through complex material interactions.

The methodological and theoretical implications of reconceptualizing the voice as an object of knowledge include considering singing, or other modes of voicing, as primarily analytical issues from the perspective of verbs rather than nouns. That is, contra views of the voice as an aesthetic, technical, or definitional catalyst, I understand voice to offer an opportunity for questioning processes that help create and perpetuate the object and idea of voice. In this understanding, assumptions about the voice as a disembodied object, or as representing a universal body, no longer gain traction.[5] By maintaining that voice, listening, sound, and music are necessarily multisensory phenomena, and by grounding my investigation in pedagogical practices—in singing and listening bodies—I not only make full use of the lessons learned in the area of sound studies, but I also open up the discipline to a broader understanding of sound by asking fundamental questions about deeply ingrained notions surrounding its focus of study.[6]

Rather than reinforcing the figure of sound, I join a current swell of work that seeks to find the nuance in and question such notions.[7] More specifically, this book seeks to recover the dynamic, multisensorial phenomenon of music and to redirect thinking about sound as object, as with the figure of sound, toward a reconception of sound as event through the practice of vibration. I undertake this project not merely as a linguistic corrective. Rather, I believe that how we think about sound matters, and that reducing a dynamic and multisensory phenomenon to a static, monodimensional one has ramifications beyond our use of the concept and metaphor of the figure of sound. My concern is that this limiting conceptualization extends to and affects all who engage with it. That is, if we reduce and limit the world we inhabit, we reduce and limit ourselves.

My claim that singing and listening are better understood as intermaterial vibrational practices may appear as a form of radical materiality, as totalizing as other metaphysical claims about voice, including voice as logos, essence, or subjectivity. However, if there is a totalizing position, it is not located within the claim to materiality. The ultimate thrust of this study does not lie in redefining and revaluing sound, music, noise, or matter but concerns those who sing and listen, and those who are moved and defined through these practices.[8] Thus, if a totalitarian position is embraced, it must lie in the relational sphere. In other words, my desire to recover the thick event is fueled by the impulse to understand more about the integral part that music plays in how we forge our relations to one another.

The Music We Name

Rather than focusing solely on a phenomenon's ontological status, Geertz advised us to examine its import. He asked: "What is it, ridicule or challenge, irony or anger, snobbery or pride, that in their occurrence and through their anger, is getting said"?[9] Reducing the thick event of music to a singular sensory mode, aurality, is driven by the high value afforded to epistemology—how to know, based on the assumption that knowing is possible—within academia and beyond.

I offer three examples. First, the requirements for knowing a given phenomenon favor particular kinds of measurements and objects that are available to be measured. In music, examples that come to mind include the fixing of pitches, the setting of tempi (for example, through metronomes), and the fascination with music that falls into the Fibonacci sequence.[10] Second, in an effort to build up areas of expertise, the drive toward adherence to the fixed referent has maintained divisions of knowledge within academia. Academic departments each claim a single perceived sense as their domain: music has claimed audition, dance covers touch and movement, art and art history focus primarily on vision (although this has changed as artists have broadly challenged the confines of that domain), and so on. Interestingly, sound, visual, and sensory studies have recently complicated these traditional domains; indeed, *Sensing Sound* is enabled by these destabilizations. Because music's agreed-on sensory domain is audition, our vocabulary and orientation are therefore primarily attuned and confined to that domain.[11] Third, academia's call to teach within these values shapes the knowledge it produces and perpetuates. Perhaps precisely because of the difficulty of knowing within these rigid confines, there is a tendency to approach the material in a mode that seems possible given the limitations inherent in its definitions.

In a radio interview, the former poet laureate Billy Collins recently described a similar disposition within the teaching and knowledge production surrounding poetry:

> It's the emphasis on interpretation, to the detriment of the less teachable, maybe even more obvious or more [sic] bodily pleasures that poetry offers. But that mental and cerebral pleasure seems to be so dominant that it leaves out other pleasures. And the other pleasures are not so teachable, so they don't require the intervention of a teacher. The pleasure of rhythm. The pleasure of sound. The pleasure of metaphor. The pleasure of imaginative travel. All these pleasures that we experience in

a gestalt fashion, you know, simultaneously as we experience a poem are difficult to discuss, really. So the emphasis tends to be on what does the poem mean?[12]

Applying Collins's insight to music scholarship and teaching, we might say that it is easier, or that it seems more scholarly, to talk about pitch, rhythm, form, historical context and debates, and meaning than it is to describe, for example, the feeling and effect of being transformed.[13] It is also easier to quantify such material than it is to convey its quality. Adherence to such values directly shapes musical discourse and teaching.

Thus we see that the analysis, interpretation, and definition of music reveal as much about ourselves (and, implicitly, about the era of which we are products) as about the music we name. That is, locating music in the musical work—which is, broadly speaking, the organization of sound—and concentrating our efforts on understanding this organization of sound might primarily yield information about an epistemological paradigm as opposed to ontology.[14] This position has been challenged. One notable example, of course, is Christopher Small's redefinition of *music* as *musicking*, a move designed to point to all people involved in music making and perceiving.[15]

The encompassing concept offered by Small's term is a model through which I begin to map the complexities of singing and listening. Similarly, the idea of transferring creative authority from composer to listener resonates with Peter Szendy's recent theory of listening as akin to "arrang[ing]" music.[16] As I have discussed elsewhere, thinking about music in this way even suggests a transfer of the privilege of authorship to the listener.[17] Furthermore, the music theorist Marion Guck put her finger on the same sore spot when she identified the false assumption that analyzing a musical work or its composer's intention alone can capture the musical experience: "As a theorist, taking listening rather than composing as an analytical focus means that who counts—the listener—is different from theory's usual orientation. What counts about the music is different, too. Since I am interested in what the listener—usually I—experience through the sounds, the point is not identifying configurations of notes but showing how my experiences are elicited by the ways in which the configurations come together for me and change me as I respond to it."[18] To advance the viability of the listener's self-inquiry as an analytical focus, we need to clarify who we are as listeners and, as such, what we can accomplish. In other words, to focus analytically on the listener allows us to read and interrogate the impact of a piece of music as it is experienced by a listener who is encultured in a given way.

Any "theory about the listener" (to invoke the subtitle from Theodor Adorno's controversial "On Popular Music") describes the results of a pedagogy arising from and representing a set of values that has produced that listening practice, rather than simply describing music lovers' "mass listening habits."[19] But it is not only in formal pedagogy (for instance, Heinrich Schenker's listening practice and that of the few composers he studied) that we can detect the underlying values that drive and direct listening perspectives today.[20] Every listening practice and its attendant theory arises from and reinforces a particular set of values.

For example, in his study of R. T. H. Laennec, who is credited with inventing the stethoscope, Jonathan Sterne observed that this technology and its allied listening practice initially developed out of restrictions, values, and attitudes related to class and gender, which called for a listening device that created physical distance between doctor and patient.[21] Jon Cruz observed that, in the abolitionist era, a listener's political position on the subjective potential of African American slaves could render the slaves' voices as either "alien noise" or "culturally expressive and performing subject[s]."[22] Both these examples speak to Mark Smith's observation that "sounds and their meanings are shaped by the cultural, economic, and political contexts in which they are produced and heard."[23] However, despite the varied nature of these observations and critiques, they all depend on one assumption that has not been fully addressed: the presumption that we can make observations, statements, and judgments about the sound of music.

In these pages I propose that sound, the narrow logic through which our concepts of music have been threaded and that lies at the center of music's definition, is merely a trope. It is an empty concept in which we have nonetheless so thoroughly invested that it has produced a kind of tunnel vision. We have taken on a stance that rejects any challenges to the a priori idea or to fixed knowledge.[24] While this assessment may be viewed as extreme, it follows from the assumption that music is a thick event. Understanding music as a figure of sound, I suggest, is merely one mode of thinking about the phenomenon. But this is an idea with enormous currency and seemingly unstoppable momentum. Not only does it shape how we discuss, conceive of, and analyze music, but it also determines the ways in which we imagine we can relate to music and the power we imagine it to wield in our lives. This shaping, in turn, influences how we configure our relationships to other humans through and with music. Indeed, the way we conceive of our relationship to music could productively be understood as an expression of how we conceive of our relationship to the world.

To be sure, in music we do experience something we call sound. However, I wish to emphasize that this is but one iteration of a phenomenon that may be defined much more deeply and broadly. While sound is a vibrational field to which we are particularly attuned, by no means does it define or limit our experience of music. Nonetheless, the conception of music as sound regularly perpetuates a host of assumptions, such as the notion that identity manifests itself through vocal timbre, a topic that I will discuss in chapter 3.

The result of the strong directing hand of the figure of sound is that when we identify and name sounds, we are not acting as free agents; instead, we are acted on. That is, because we have allowed music discourse to rely so strongly on the figure of sound, it pulls us toward certain ways of experiencing and naming sound and limits our access to other ways. As a consequence, we are not entirely free to experience sound idiosyncratically or to experiment unrestrictedly with that experience beyond agreed-on names and meanings. In fact, if such unbounded naming were carried out, the resulting definition of not only music but also sound itself might not fall under conventional notions of sound. For example, a given phenomenon is, under the figure of sound, understood as the spoken sound /b/ or /p/. In contrast, when released from the figure of sound, the same phenomenon may be understood as an event that, because of the amount of air it emits, has a greater or lesser impact on the skin.[25] Indeed, if the naming of a given phenomenon were uncoupled from the logic of the figure of sound, parameters that currently define this suite of phenomena might be considered not as fundamental, but as merely marginal.

My project arose from frustration with the ways in which, in contemporary musical discourse, we fall short in thinking and talking about (and in devising and interrogating performative and listening practices around) sound by relying largely on judgments about meaning and morality (for example, "she listens well" and "he listens poorly").[26] By critically assessing notions of sound as perceived through the lens of a meaning-making or sound-making source, I try to capture the ways in which a vibrational force is reduced to statements like "this is the sound of a trumpet" or "this is the sound of a black man," and I attempt to broaden such perspectives. Thus, beyond this volume, I envision a move toward analytical models that simply and elegantly challenge such reductions and their impacts.

Were *Sensing Sound* a historical study, my task would be to directly address how the vibrational material phenomenon, as I understand it, has been conceptualized, understood, and acted on in disparate geographical and historical contexts. While that undertaking would be fascinating, and perhaps one for a future date, what I offer here is rather a contribution to the contemporary debate, in light of recent currents in opera, sound, and sensory studies concerning how to conceptualize and analyze some of the music that is performed and heard today by contemporary artists and audiences.[27]

Sensing Sound rejects the position that sound is a fixed entity and the idea that perceiving sounds depends on what we traditionally refer as the aural mode. This rejection triggers two pivotal questions. First, is the listener's or musician's awareness of and/or sensitivity to these multisensory sensations essential to this rejection and to a possible alternative position? (A related question is, would my argument need adjustment depending on the answer to this question?) Second, does my reframing of sound apply only to the particular and extreme repertoire treated here? For me, the answer to both of these questions is a resounding *no*! The observations gathered here reveal that, indeed, most people are unaware of the sensations or modes of what we refer to as sound and music. Common musical discourses tend to steer perception and analysis toward particular experiences — especially toward the auditory mode. I do not, however, invoke a Cageian move toward listening to all sounds, including the sound of silence, and the aesthetics of panaurality.[28] On the contrary, I maintain that not only aurality but also tactile, spatial, physical, material, and vibrational sensations are at the core of all music. Because the figure of sound produces a listening practice and a subject position that can perceive only within that mode, it is challenging to imagine anything outside it. Therefore, it is within these limits that I found my case studies.

Music's Naturalized Cornerstones

Given that the fundamental concepts and vocabulary which we use routinely in making sense of music are thoroughly naturalized, how can we possibly think and experience beyond them? The performance studies theorist José Esteban Muñoz introduced a useful analytical tool for envisioning ways in which the essentialized body and, by extension, the essentialized voice may rewrite or decode itself. This model has been useful in my efforts to think about extraparadigmatic experience. Building on the cultural theorist Stuart Hall's encoding or decoding modes, Muñoz defined "disidentification" as "a hermeneutic, a process of production, and a mode of performance."[29] Muñoz likened disidentification to what Hall defines as the third and final mode of decoding, in which meanings are unpacked for the purpose of dismantling dominant codes to resist, demystify, and deconstruct readings suggested by the dominant culture — that is, as an oppositional reception. Disidentification, according to Muñoz, is an "ambivalent modality," the minority spectator's survival strategy that "resist[s] and confound[s] socially prescriptive patterns of identification."[30]

Disidentification, which Muñoz exemplified through readings of drag performances with explicit racial references, is thus a performative stance undertaken with deep knowledge of essentialized subject positions. Through the rewriting, decoding, or double performance of such subject positions, the unspoken values that provide the contours, akin to unerased text, may surface; quotation marks appear around the essentialized subject position. Through purposeful foregrounding of the text layered through a series of rewritings, these meanings no longer simply hover in the background, passively confirming what was thought to be the subject's essential truth. Instead they are materialized and externalized, and through this process we are finally able to acknowledge them. Moreover, it is by first acknowledging the overarching a priori framework through which the world is comprehended that we can recognize both essentialized subject positions and naturalized notions of sound, and their mutually reinforcing effects.

While I am indebted to Hall's and Muñoz's powerful work, I also recognize that their interventions (like most scholarship on race) remain within an orbit wherein signs and signifieds are relied on in social transactions. In essence, they critique the power and effects of signs when used or interpreted unjustly. However, both the critique and the solution they provide are spun from, and limited to, the figure of sound's centrifugal logic. And it is with this logic—instrumentalized through its agreed-on parameters—that music's naturalized cornerstones are laid and cemented. The figure of sound has been so thoroughly naturalized that our belief in its certainty is akin to our reliance on gravitational force.

I hope that this book will offer a convincing "yes" to a vibrational theory of music (and to a subsumption of sound under vibration) and to an alternative analytical framework to that offered by the figure of sound. In grappling with contemporary vocal performances that do not yield to analytical frameworks premised on the figure of sound, I was emboldened to think about naturalized notions in music in new ways. Rather than rejecting them as nonsensical, which was admittedly my first instinct, I needed to allow the performances themselves to show me how to approach them. The performances had proved unyielding to familiar analytical frameworks not because they had failed in an a priori way, but because those techniques of analysis available to me had been created to understand particular music—music built on a different premise than the performances I had at hand.

Viewing music in this way carries some unsettling consequences. First, it suggests that traditional approaches constrain our understanding rather than expanding it. Second, it asks that people who interact with, are touched by, and seek to understand music approach an artificially bounded experience without that familiar scaffolding. It asks anyone seeking to understand music to let go of the safety net of assumed certainty that is offered by reliance on musical parameters and concepts, and instead to enter the apparent chaos that follows the rejection of preconceived categories.

If this was the sole effect of a vibrational theory of music, its disruptions would be destructive. But approaching music as a vibrational practice offers much more: it recognizes, and hence encourages, idiosyncratic experiences of and with music. Furthermore, approaching music in this way takes into account its nonfixity and recognizes that it always comes into being through an unfolding and dynamic material set of relations.

Therefore, though unsettling at first, augmenting or replacing fixed musical categories (and their attendant parameters, endowed with value by a given culturally and historically specific situation) offers an opening. It enables us to recognize our interaction with and participation in music, and our interaction with and participation in the world, in ways that we have always intuitively recognized and always strongly felt, but that we were seldom empowered (or encouraged) to articulate.

It bears mentioning that a license to take the materially and vibrationally specific experience—the thick event—as a starting point is the opposite of self-centeredness. Taking vibrational practice as a basis for knowledge building around music's ontology and epistemology turns our attention from the categorical correctness or incorrectness of a given description of music to the ever-changing relations that constitute music. As in deconstruction's signifying chain, the final meaning in vibrational practice is endlessly deferred. Moreover, by recognizing vibrational practice or the thick event as ground zero, we are reminded to note and articulate our experiences of music in ways that always keep in sight, and in ear, the ethical dimensions of sound, music, singing, and listening.[31]

To fairly consider the performances at hand, I engaged themes both central and peripheral to the musicological debate. As a result, by adding multisensory and material considerations to the powerful and effective work of Hall, Muñoz, and others, I approach what we have traditionally conceived as sound from six interrelated transdisciplinary concerns: the body, the sensory complex, the sound, the (performative and experiential) methodological orientation, the analytical orientation, and the metaphysical.

I approach the body in and as performance, and as it manifests itself to us as a result of cultural construction and habituation. I consider the sensory complex of voice, sound, and music with similar mindful attention to the ways in which that complex by definition is culturally structured. And I keep in mind that any information we might glean through the sensory complex is thus shaped. This perspective leads me to interrogate the culturally informed parameters of sound on which we rely. That is, does any music exist prior to and independent of that which a culturally structured and informed sensory complex gives rise to, delivers, and verifies? Or—as the question of the falling tree's sound suggests—is the music we can sense in any given cultural moment merely a reflection (or indeed a confirmation) of our limited ability to perceive that moment?[32] The process of responding to these questions led me to interrogate musicological cornerstones: musical parameters, methodologies, and analysis.

I also interrogate one of music's fundamental parameters: sound. I do this because the traditional understanding isolates sound from the thick event of music—a parameter from which we believe we can derive knowledge of music and its effects. In so doing, I retreat from the assumption that music lies uniquely in the sphere of sound. Taking that assumption seriously, I pay close attention to the gradations and impacts of vibration (as in sound), transmission (as in intermaterial flow), and transduction (as in conversion of wave form from, say, mechanical to electric) within historical and theoretical discourse. My study relies on a methodological orientation which arose from a concern that I was trapped within my vocal training's culturally and historically shaped and informed perceptual structures. Hence my methodological orientation includes attempts to disrupt said sensory complex by working through vocal and listening practices that explicitly refuse to concern themselves with sound making or conventional aural-oriented listening. Moreover, I turn my attention to the question and issue of analysis, specifically to self-consciously interrogating where we direct our analytical focus and with which methods we decipher our material. I also note that the metaphysical assumptions at the base of musical inquiry arise in relation to questions about music's materiality or ineffability. Finally, I should mention that, as my references to Hall and Muñoz have suggested, my grounding orientation is informed by some of the critical perspectives and insights offered by scholarship on race and gender.[33]

My methodological orientation, then, is based on the premises that, on the one hand, dominant concepts are (silently) instilled in the human body and that, on the other hand, by testing a concept through its use in teaching, the concept's (unintended) consequences may be revealed. By following singers who sing in ways or locations that do not fit into the dominant concepts of singing, we can begin to sense the outlines of these dominant concepts—which, precisely because of their dominance, are naturalized under more normal circumstances, and hence are beyond the purview of our critical and analytical focus.

Thus I investigate underwater singing and singing that does not engage the vocal cords, in both theoretical and participatory modes. To interrogate the possible connections between the practice of singing and the concept of the figure of sound, I follow that concept into the vocal instruction studio. In doing so I can ask: When we use the concept of the figure of sound, how does a body that is poised to make sounds react? Furthermore, what does the result tell us about the viability of the concept? I can also play with, and test, other concepts of voice and sound. The comparative results are concrete, presented in terms of how a voice student feels and performs based on the two types of instruction.

I build on scholarship that has made great strides toward a thorough consideration of the body's role in musical experience.[34] To summarize, I think about this work as having two variants that attempt to accomplish separate yet interrelated goals. One variant mines the body as a site for valuable information regarding the composition or performance situation and how the corporeal cultural formation and general environment (what is allowed and not allowed in terms of the body) informs what seems available as compositional and performative possibilities. Another variant largely consists of work by scholars who were trained outside musicology, but who are nevertheless serious scholars of sound. The latter considers how the full spectrum of sensory experience contributes to our interpretation of sound and music. Less has been done in this area of research to address the musical repertoire in particular.[35]

I have found it useful to think about the body within the realm of sensory studies and material scholarship. To me, this perspective removes perceived barriers between music scholarship and the sciences and medicine. It does not distinguish between production and perception but sees them as creating each other. The title of Jody Kreiman's and Diana Sidtis's groundbreaking book, *Foundations of Voice Studies: An Interdisciplinary Approach to Voice Production and Perception*, articulates this cocreating dynamic. The authors recognize that the analytical object that comes into relief is a direct consequence of the way in which it is processed by our culturally formed sensory complex. Consequently, an analysis of voice cannot concern only the so-called object but must also include the process that defines and recognizes it as such. Thus, the sensory and the material go hand in hand. Expanding our tool kit of perspectives to include select aspects of what the sciences and medicine can offer moves us closer to understanding voice, sound, and music and the sense we make of them.

A major aspiration for this project is to suggest a framework for, and offer an example of, analysis of voice and music that takes its analytical cues from the vocal and musical event at hand, rather than from a music-analytical framework developed with a particular repertoire (and different goals) in mind.[36] Applying these interlocking and mutually fulfilling perspectives, I take inspiration from scholars who engage in microhistories (that is, in-depth historical work on limited repertoires), and I adapt such a detailed approach to a close analysis of previously excluded factors. Hence, my analytical orientation takes the form of extending methods and strategies from sound studies and sensory studies and applying them to issues arising in contemporary opera studies, contemporary music, and the emerging discipline of voice studies. Examining aspects of the vocal or musical event beyond the normalized parameters of traditional music analysis, I extend perspectives offered by sound and sensory studies to the multivalent, simultaneous, nuanced processes and effects of lived music. When I consider the shared sensory activities of singing and listening, my emphasis is on microanalysis.

This level of analysis shifts the focus on music to a finer-grained level than that of pitch, rhythm, form, and other commonly considered musical parameters, and I find that this approach resonates with aspects of Carolyn Abbate's work. Drawing on Vladimir Jankélévitch, Abbate argues that "music's effects upon performers and listeners can be devastating, physically brutal, mysterious, erotic, moving, boring, pleasing, enervating, or uncomfortable, generally embarrassing, subjective, and resistant to the gnostic."[37] In other words, our actual experience with music is experienced rather than reasoned and interpreted; "drastic," rather than "gnostic." However, my response to the drastic versus gnostic dilemma to which she calls attention is, first, to develop a critical framework for dealing with the so-called drastic aspects, especially one that seeks to tease out the naturalized notions through which we understand sound.[38] Second, I argue explicitly that we can—in fact, we have a responsibility to—attempt to understand the drastic in organized analytical terms, and indeed in its entanglement with the terms set by the gnostic.

In so doing, I draw on models developed by scholars who traverse the terrain of music, sound, technology, media, and the senses. For example, Martha Feldman's work on the castrato voice and Emily Dolan's work on orchestral timbre have already begun forging lines of inquiry about the coupling of shifting aesthetic sensibilities with the onset of new technologies, medical or otherwise.[39] And scholars working on issues of technology and disability have, by necessity, had to consider the intersection of dominant material structures of perception and technological invention.

Mara Mills's historical work on the question of media, the telephone, and deaf culture cannot but tell a story about the perceived limits and ideals of the sensory complex, and about the material implements created to bridge such imagined shortcomings.[40] Veit Erlmann's historical work on modern aurality suggests that, historically, a particular type of epistemology has defined reason in direct opposition to resonance.[41] Along the same lines is Joseph Auner's work on musical modernism in the first half of the twentieth century, as marked by the sensitivity of the "phonometrograph"—Eric Satie's term for "weigh[ing] and measur[ing]"—that is, modernist sensibilities indelibly created by "ears and minds remade by recording, phonography, player pianos, and the burgeoning science of sound."[42] Furthermore, Alain Corbin's influential work on nineteenth-century French village bells and the ways in which their physicality (including patrons' inscriptions) and sonic reach was an intimate part of villagers' interpretation of their sound has been a crucial model of a powerful analysis.[43]

Building on these and additional important perspectives from disability and media studies, history, and musicology, my approach differs from the majority of items in the current onslaught of work by new materialists in that I take a stance on the lived material body, and that my primary motivation is to learn about the material relational dynamics gleaned from feminist and race studies.[44] But, when I lean toward a material approach that takes into account material's vibration, I take my strongest cues from scholars such as Elisabeth Le Guin, with her dedication to "cello-and-bow thinking," James Davies's "avowedly *realist*" stance on the question of how "music acts in the cultivation of bodies," and Peter Lunenfeld's commitment to "maker's discourse" when thinking through digital and media practices.[45] My perspective and motivation are informed by my practice as a classically trained singer who has worked in close musical collaboration with composers as well as in improvisational settings. My thinking has also been informed by the contradictory ways my voice has been read, depending on whether the listener has access to visual (Korean) or sonic (Scandinavian accent) cues. Furthermore, my many years of learning about voice and listening to voice as a voice teacher have left indelible imprints on my theoretical orientation. In my experience, nothing forces me to come to clarity about a given topic, concept, or practice like having to articulate it in teaching.

Additionally, given that most of the vocal apparatus is hidden from the naked eye and that most vocal mechanisms are comprised of involuntary functions also used for basic survival (such as breathing), teaching voice is a notoriously elusive and challenging craft.[46] Hence, echoing the saying, you learn what you teach, my litmus test in regard to my knowledge about voice is whether or not, as a voice teacher, I can help a person use his or her voice in a way that person would like to. In large part, what I know about voice and listening, and what I employ in my theorizing, is drawn directly from this experimental and experiential practice.[47] Therefore, while the position communicated herein is in intimate dialogue with and irreversibly influenced by theoretical perspectives, it has first and foremost been developed through my experience as a teacher and student of voice, and as a student of listening and human relations. I think about this through the Norwegian term, *håndarbeid* (meaning the work of the hand) — a practice and concept that can broadly be translated as the domain of doing.

Finally, the entirely unintended theoretical implications of this project result in a strong position vis-à-vis the metaphysics of music. In this way, I partake in the conversation begun in the 1980s when musicology underwent a tectonic shift with the onset of scholarship that self-consciously sought to inquire beyond positivistic values into music. In Susan McClary's words, positivistic scholarship was limited in its understanding music as "a medium that participates in social formation by influencing the ways we perceive our feelings, our bodies, our desires, our very subjectivities — even if it does so surreptitiously, without most of us knowing how."[48] Integral to that new conversation was Small's notion of "musicking," a concept that has become key to analyses of musical life and that, as mentioned earlier, has influenced my own thinking tremendously.

Learning from Small and others, we might think about the question of the falling tree by considering the community that planted the forest and that community's needs and hopes for that plot of land and what it yields. We might consider too the dynamics among the different social, cultural, and economic circumstances represented by the people who come together around the land — for example, farm workers in relation to forest rangers, and forest rangers in relation to those using the forest for recreation. We might ask questions about their varying aspirations and their social and aesthetic needs and desires. New musicology's perspective offers invaluable access to social, class, cultural, gendered, and economic dynamics.

Small's project of rethinking the social dynamics of music through the concept of musicking may have its parallel in thinking about music and sound as the transmission of energy through and across material. While Small expanded the discussion from music as a "thing" to music as an "activity, something that people do," including perspectives from sound, sensory, and material studies, I pay attention to the microscopic material transformations that music helps to usher into reality.[49] And as Small's definition of music put the social at the hub, I hope that this discussion can expand the conversation further, from thinking about music as a knowable aesthetic object to thinking about it as transferable energy.[50] *Transferable energy* here denotes energy pulsating through and across material and transforming as it adapts to and takes on various material qualities; it is at the crux of thinking about music in the dimensions of nodes of transmission and vibrational realizations in material-specific and dynamic contexts.

Situated within musicology and its intellectual trajectory, I have found that the concept of vibration, considered in a musical context, is useful when putting cross-disciplinary bodies of knowledge in dialogue.[51] While the concept of the figure of sound represents a disregarding of areas of knowledge that fail to fit within prescribed frameworks, vibration provides a route for thinking about fluidity and distribution that does not distinguish between or across media, and a portal for communicating beyond physical boundaries. For example, the political scientist Jane Bennett relied on an obscure treatise on music in developing her arguments for the "political ecology of things" and the "active participation of non-human forces in events."[52] Toward that end, she theorized a "vital materiality" running through and across bodies, both human and nonhuman.[53] Like Bennett, I am concerned with the material relationship between humans and things, for which the practice of vibration is both metaphor and concrete manifestation. And I see music not as a novel example of vibration, but as an everyday example of that tangible, material relationship, akin to tree leaves' movements manifesting the wind.

Music as Nodes in a Chain of Transmission and Transduction

Thinking about music through the practice of vibration brings up the limitations of the paradigm of music as sound, as articulated by Rebecca Lippman, a participant in one of my graduate seminars: "But if we think about this phenomenon as vibration, where does vibration begin and where does it end?"[54] With this question, Lippman encapsulated the limitations of our conceptualization of music when we operate with naturalized notions: the set of questions and observations central—perhaps native—to one paradigm often seem foreign and irrelevant to another. For example, within one paradigm we would consider a certain phenomenon to be sound and see it as bounded and knowable, with a distinct beginning and end. Yet within a different paradigm we would see the same phenomenon as vibration and understand it in the terms of the energy in a body's mass and its transmission, transduction, and transformation through different materials.⁵⁵ Furthermore, while the first paradigm includes parameters, such as duration, that specifically imply beginnings and endings, these parameters—duration, in particular—are less relevant in the second framework. Within that framework, relevant information comes from inquiries into the relationships between materials and sensations, indeed between the bodies involved. Each paradigm has its own logic, and the parameters and questions that yield knowledge in one are not necessarily productive in the other. Let's compare the two frameworks:

Figure of sound	Practice of vibration
— Remains the same independent of listener (fixed)	— Shifts according to listener (relational)
— Circumscribed	— Always present
— Defined a priori	— No a priori definition
— Original; copy	— No assumed original; no copy
— Judged according to fidelity to source	— Nodes of transmission observed
— Static	— Dynamic

The figure of sound is an entity whose existence depends on an objective measurement. For instance, sound as a figure demands a concrete definition on a larger scale of bounded territory, as does the ground in a figure-ground relationship. If the smaller scale is, for example, pitch, the bounded territory is song. Vibrations, however, are unbounded: their relations are defined by process, articulation, and change across material. In this paradigm, then, the phenomena that we conventionally recognize as notes making up songs cannot be limited to particular renditions or articulations. What we observe and label as sounds in the figure of sound framework are considered simply as different points of transmissions in the practice of vibration framework. If singing and listening both constitute the process of vibration across material, they are always present—or, more correctly, always occurring. In short, listening to, making, and manifesting music is a vibrational practice.

From the perspective of this practice, it is the impetus, the urge, and the rush to action—indeed, the vibrations that this presonic activity puts forth—that make up singing and music making. In other words, sound is created and shaped in the action and transmission of vibration, millisecond to millisecond. A person's body is also conditioned, shaped, and created within that timeframe, and the sounds it can produce are determined—and limited only—by the range of action and material transmission. That is, we participate in the points of transmission: for each of us, there is no knowable music or sound before its singular transmission through us. While each iteration is unique, we exist as a sine qua non, and the vibrational energy exists prior to the particular transmission.

This completely contradicts the figure of sound's drive to define sound according to an original, and to apply the question of fidelity to a source. Furthermore, without a drive to identify an object, or sound bounded by a beginning and an end, there is no assumed original with which to compare and against which to measure a given figure of sound's relationship and potential legitimacy. The evaluation of fidelity assumes a static object, which is examined to determine its relative loyalty and similarity to the source; in contrast, the practice of vibration assumes a dynamic, shifting process of transmission.[56] In other words, when there is no assumed fixed object, the need to establish relative fidelity to a static definition evaporates.

As Lippman's question reveals, the figure of sound paradigm assumes that knowable and measurable things form the basis of music. A considerable amount of music analysis derives its main energy from defining these objective elements and naming their relationships and structures. While we understand that defining pitches within scalar systems is contextually dependent within a particular discourse about a musical system, we accept that a given analysis and its attendant listening practice and judgment do not question the basic building blocks of the analysis (for example, pitch). Within the sound paradigm, a given pitch operates as a stable index or signifier. While a range of values and beliefs is tied to the signifier's assumed relation to a given sound, this framework impels us toward recognizing a given iteration's fixed relationship a priori.[57]

This plays out dramatically in music: a given epistemic framework developed through a cultural system enables us to recognize and name, say, a G#. In other words, G# is historically situated within a chromatic, tempered scalar system that is culturally bound to the Western tonal system. Recognizing the vibration that we name G# also assumes recognition of the system within which G# is situated, including a number of possible systems—for instance, the assumption that it is part of the E-major scale but that it would be a foreign note (indeed, the tritone) in a D-major scale. Recognizing G# also leaves out the possibility that these vibrations play a part in other musical systems that would not recognize them as G#.

However, the paradigm of the figure of sound does not stop with the drive to know and identify a pitched sound as the second scale degree of F# major: it is bound up in the assumed meaning of this identity, and it is often derived from values and assumptions about identity that are deciphered from visual clues.[58] The figure of sound paradigm so structures listening to voices that it can lead to appraisals such as "this is the sound of a woman's voice." This appraisal is based on perceived similarities and dissimiliarities between one sound and another — in this case, on similarities to other human vocal sounds and on dissimilarities to, specifically, men's and children's voices.[59] By assuming an essential tie between a vocal timbre and a given definition of race, this paradigm can also lead to observations that are loaded with a presumption, such as the voice "sounded as if it was of a male black."[60] Listening to voices through the framework of sound can also carry multiple layers of appraisal: for example, the observation that somebody is "talk[ing] white."[61] This judgment has at least two layers: the idea of "talking white" assumes that the speaker is not white, and that the unexpected racialized vocal style is relevant only because of that assumption. (Just as the designation G# can be applied in relation to many different scale systems, the observation that a person is "talking white" can be applied against a backdrop of a number of different racial classification systems.)

Ultimately, the figure of sound reduces sound's being and its attendant listening practices to sound's relative relation to a range of a priori ideas of sound. It also reduces the listener. In this dynamic, the listener's main task is to name the relationship between figure and ground: the task revolves around determining a sound's faithfulness to a given set of assumptions. Here, being faithful entails such virtues as being in tune and conveying the a priori intent and meaning of a particular sound, composition, or musical-cultural tradition. From the assumption of a defined, nameable, and knowable sound follows an assumption of fidelity, and a perceived moral obligation to consider each sound in its fidelity to that a priori. Robert Fink aptly describes these two processes as "listening through" a sound versus "listening to" that sound (for itself).[62] In other words, this model rests on the assumption that, in the meeting between a sound, a voice, and a music, the respectful, responsible, and ethical way to relate to the sound, voice, or music is through the capacity to recognize it and know it.

The practice of vibration, in contrast, relates a sound not to an a priori definition but to transmission. Because propagation is never static and, as a series of continually unfolding transmissions, is not a matter of recognition and naming, the notion of fidelity accompanying the figure of sound is undermined. If there is nothing to which sound must remain loyal, the notion of fidelity does not retain its currency. Then, rather than limiting our conception of singing to the task of replicating an ideal sound, we might grow comfortable with the notion that human existence and the activity that flows from a human being necessarily constitute a song. Singing beyond the "shadow" of the figure of sound then moves away from forcing us to mold our bodies to create an expected sound, and toward accepting the vibrations that pulsate from our material, sonorous beings.[63]

Before discussing the larger ramification of this modulation from the figure of sound to the practice of vibration, I should stress that I do not elevate vibration merely in an effort to move away from a perceived linguistic hegemony based on the figure of sound. My approach to the consideration of music as a practice of vibration is not just a definitional adjustment, nor simply a rhetorical attempt to allude to prelinguistic and presemiotic spaces or pre- and posthistorical spaces. In invoking vibration, I am not making a posthuman move toward the subjectivity and agency of things, or away from human-made sounds to theoretical vibrations of the spheres, unrelated to and unencumbered by humans. I reach toward vibration not to offer a mechanical orientation or to align considerations of sound with science, nor because I consider music as entirely mechanistic, something in the sphere of applied engineering rather than aesthetics.

Instead, my turning to vibration is fueled by my interest in thinking about music as practice, not object. Music as vibration is something that crosses, is affected by, and takes its character from any materiality, and because it shows us interconnectedness in material terms, it also shows us that we cannot exist merely as singular individuals. In this sense, music as vibration is analogous to social relations in a Marxist sense, or "the common good," which, as the theologian Jim Wallis cites from Catholic teaching, is vital to the "whole network of social conditions which enable human individuals and groups to flourish and live a fully genuinely human life."[64] The ramifications of understanding music as a practice of vibration are not limited to music discourse or music culture, as Wallis has suggested. In contrast to the figure of sound, the figure of vibration understands music as always coming into being: it renders music an event of the common good.[65]

This shift in orientation leads to major adjustments regarding epistemology, ontology, and ethics. First, using the illuminating framework of the Dutch philosopher and anthropologist Annemarie Mol, "ontology is not given in the order of things, but . . . instead, ontologies are brought into being, sustained, or allowed to wither away in common, day-to-day, sociomaterial practices."[66] Second, when we deal with music, singing, and listening as events rather than as objects, the need for a specialized epistemology of sound evaporates. Questions and methodologies designed to lead to the ability to know and identify the sonically knowable become uninteresting if there is nothing to recognize and identify a priori, nothing to know.

And, third, this epistemological shift replaces the central tenets of musical ethics and values, moving from fidelity (questions of identity and difference) to charity (concern for the material implications of our actions on others). Here, we consider the experience of music as one possible register in the full range of material vibrational practice. If we accept this position, music necessarily brings us into the territory of relationality, and hence of political ontology. Thus, what we conventionally consider audile listening is only one of many possible ways of articulating and interacting with and through material relations.

Naturally, then, music is only one of many areas in which adopting the paradigm of the practice of vibration helps both equalize the roles and contributions of the different senses and point to an ethics that circumvents fidelity. For example, a thought model that I have followed, and that has influenced me throughout this project, is Aldo Leopold's classic essay "Land Ethic," first published in 1949.[67] In it, and through his lifework, Leopold introduced ethics as the fundamental concept that should underlie all considerations of land and water use, including our relationship to land and water. While my project does not explicitly argue for sound making and listening as ecological practices, I have found in Leopold's philosophy of the human-land relationship a lucid model for human-human relationships as they are rendered when sound is understood as material transmission: "In short, a land ethic changes the role of Homo Sapiens from conqueror of the land community, to plain member and citizen of it. . . . It implies respect for his fellow members, and also respect for the community as such."[68] Leopold's text, which is intensely relevant today, is valuable in thinking about all relationships and stewardships into which humans enter. While reading the above excerpt, in my mind's ear I heard: "Approaching sound, music, and voices as vibrational practice changes the role of Homo Sapiens from conqueror of the figure of sound, to plain member and transmitter of a vibrational field. It implies respect for his fellow members, and also respect for the community as such."

Leopold's meditation on our ethical relationship to the land resonates with and underscores my convictions about ethical relations in the practice of music. Trapping music in the limited definition that follows from the figure of sound (that is, a stable signifier pointing to a static signified) constitutes an unethical relationship to music. According to my definition, having an ethical relationship to music means recognizing it as an always becoming field of vibration and realizing that music consists not only of inanimate materials, but also of the materiality that is the human body. Starting from Leopold's clear vision about the human-land relationship and adapting it to human-human relationship with an understanding of music as material transmission lays bare how we are interconnected: "It's inconceivable to me . . . that an ethical relationship to [music] can exist without love, respect and admiration, and a high regard for [human] value."[69]

Leopold reminded us that we do not possess the land; rather, we have been entrusted with its stewardship.[70] Similarly, because a sound cannot be fixed, one cannot own a sound. In our relationship to sound we are both in and of vibrations. We simultaneously create and experience vibrations, sound, and music in the same moment, both as performers and as listeners. And it is precisely because vibrations do not exist separately from the materiality of the human body that we cannot objectify them.[71] Sound, voices, music, and vibration are under our stewardship as long as we are part of their field of transmission.

Chapter Overview

My denaturalization of music's parameters and investigation into music as a vibrational practice unfolds over five chapters. Four of these chapters use twenty-first-century American operas—envisioned and created by a rich range of women composers and performers—to think through four naturalized ideas about singing, listening, sound, and music that commonly underlie musical perceptions and discourses:

- The privileging of air, as opposed to any other medium of sound propagation;
- The predominant idea that sound's behavior should be understood in linear, visual terms;
- The presumption that sound is stable, knowable, and defined a priori; and
- The assumption that music deals only in sound and silence.

Each of these naturalized ideas typifies a flattening of what I posit is a multidimensional and contextually dependent phenomenon. And each depends on a priori definitions of sound.

In the first four chapters, I denaturalize these presumptions, which are the bedrock of many musical analyses and colloquial conceptions. These case studies arise from my engagement with multisensory scholarship, sound studies, voice studies, and opera studies. I generalize this analytical framework in the book's final chapter, considering music as a vibrational event and practice. In pursuing this line of inquiry I come to the understanding that, because music is not apart from us but of us, it cannot be naturalized. Hence my concluding chapter makes it clear that my critique of fundamental sonic conceptions is indeed a critique of their ethical implications.

In chapter 1, "Music's Material Dependency: What Underwater Opera Can Tell Us about Odysseus's Ears," I examine the underwater vocal practice of the Los Angeles–based performance artist and soprano Juliana Snapper (b. 1972) and dispense with the idea that sound is stable and knowable before it is produced and perceived. By no longer viewing air as the natural medium through which sound materializes, and by recognizing instead that airborne sound partakes of air's distinctive features, we come to appreciate the process of sound as a dynamic, interactive coming into being. This chapter also applies Snapper's insights to a surprising new reading of the sirens in Homer's *Odyssey*. This is the first of three chapters that discourage the common understanding of sound as merely aural and expose the associated deficiencies in current analytical techniques.

In Chapter 2, "The Acoustic Mediation of Voice, Self, and Others," I deal with spatial-relational and acoustic dimensions that are naturalized through distinct sonic, performative, and listening practices. The two pieces I examine, Meredith Monk's (b. 1942) 2008 *Songs of Ascension* (originally composed for a sculptural tower with a double helix stairway and subsequently rearranged for traditional performance venues), and the opera-for-headphones production of Christopher Cerrone's (b. 1984) 2013 *Invisible Cities* (performed within the bustle and everyday activity of Los Angeles's Union Station but delivered to audiences via headphones), show that most of the live music we hear in a Western context is presented within an acoustic frame so naturalized that any other acoustic setting is understood as wrong rather than different. I suggest that a given acoustic frame offers us more than simply poor or optimal sound, and that thus the naturalization of acoustics affects dimensions beyond our experience of the sound per se. That is, I posit that acoustic and spatial specificity also participate in giving form to the figure of sound, and that the acoustic mediation of sound and habituations related to it profoundly influence our experience of self and others.

In Chapter 3, "Music as Action: Singing Happens before Sound," I posit that sound is a subset of vibration and suggest that singing and listening are vital exchanges of energy. I interrogate the basic principles of singing and sound production by examining performance art pieces by Elodie Blanchard (b. 1976) and a chamber opera by Alba Fernanda Triana (b. 1972). In these projects, sounds do not maintain static definitions based on numerical values (for example, 440 Hz) or significations (such as the note A). Instead, sound is a dynamic element arising throughout the exchange that takes place during singing and listening. This chapter denaturalizes sign- and discourse-based analyses of sound, proposing in their place a material, sensory-based analysis that assumes sound to be the result of an action rather than the action itself. I compare this perspectival shift to the sea change that took place in art criticism in response to Jackson Pollock's work: with the rise of what became known as action painting, critics had to move away from defining artistic work as a corpus of reified objects (works) and instead define it in terms of the actions that might have produced such objects. In this way, chapter 3 questions the position and origin of the definition of *work*.

Chapter 4, "All Voice, All Ears: From the Figure of Sound to the Practice of Music" concerns common assumptions about music and its definition. One major problem with the naming process in general is that the name becomes an index for an experiential phenomenon. Relying on the index, we become several steps removed from the phenomenon itself, including its initial, singular articulation; the likelihood that we can experience another moment unmediated by prescribed parameters and meanings; and even the name itself. For example, although we are educated to believe that it is the form of an opera that moves us, in actuality we are moved by multiple singular and particular articulations within, yet not reliant on, the operatic form. We listen for opera, arias, and a particular operatic sonority; we endorse and validate the experiences we have in accordance with these predetermined categories at the expense of other experiences—that is, even though other articulations that do not fit the categories might also offer meaningful experiences. Thus the names, and the fit between names and experiences, become central. This constitutes the process of reification. In chapter 4, I examine how this process is performed in classical vocal pedagogy, and I experiment with a teaching style predicated on the assumption that singing and music are material articulatory processes. This chapter proposes that articulatory action—indeed, events—is at the core of both singing and music.[72]

The fifth and final chapter, "Music as a Vibrational Practice: Singing and Listening as Everything and Nothing," uses the four case studies and multisensory perspectives offered by the preceding chapters to propose a model for thinking through selfhood and community. In this model, *we* are sound. Like sound, which comes into being through its material transmission, human beings are not stable and knowable prior to entering into a relationship; rather, we unfold and bring each other into being through relationships. Our potential for recognizing and accepting self and other rests on our ability and willingness to be changed by our encounters, rather than merely by the potentially desirable qualities (or their absence) in others. Hence, for a relationship with sound to take place, we must be willing to take part in, propagate, transmit, and—in some cases—transduce its vibrations. From this it follows that entropy occurs when we focus on the preconceived identity of another rather than on our own ability (or inability) to undergo change. I posit, then, a strong parallel between how sound is realized or propagated through certain materialities and how we as unique beings are being realized through transmission and the reception of another person who approaches us as a unique, unrepeatable human being.[73]

MUSIC'S MATERIAL DEPENDENCY
What Underwater Opera Can Tell Us about Odysseus's Ears

In space, nobody can hear you scream. —ALIEN (1979)

Prelude

In 2007, I received an invitation to a recital that would take place in my bathroom. The Los Angeles–based soprano and performance artist Juliana Snapper, a soprano and performance artist who works in experimental music, offered to come to my home and present an underwater concert in my tub. "Crazy," I thought. "Why go to the trouble of singing in an element so far from ideal?" Why choose a setting that would conjure up clichés from the Homeric sirens to Disney's *The Little Mermaid*? I declined the invitation, yet Snapper's endeavor lingered in my thoughts.

Introduction

Fundamental to Western notions of sound and music is the assumption that we can know and recognize musical parameters. Indeed, these notions form the basis of Western music's classical analytical practice and, as I suggest throughout this book, of the culture's quotidian "audile techniques," to evoke Jonathan Sterne's useful term.[1] Even in contemporary works and studies, where traditional scores may not be relevant, dependence on sound waves, timelines,

FIGURE 1.1 · Juliana Snapper at the Aksioma Institute for Contemprary Art, Ljubljana, Slovenia, June 20, 2008 (photo by Miha Fras; courtesy Aksioma, Ljublkana).

and algorithms maintains the traditional tendencies to quantify music. Consequently, an abstract yet fixed notation, or a notation-derived notion of sound, overshadows the actual, ever-shifting experience of music. In vocal studies, this orientation plays out as a privileging of dramatic, structural, and semiotic content derived from documents (libretto, score, and contemporaneous documents) and analyzed with attention to the sociohistorical context over the distinct quality or timbre of each individual voice in each performance of each work. Historically, Western music studies have favored the idealized and abstract at the expense of the sensible, unrepeatable experience.

The common conception of the voice as a generic vehicle for words, pitches, and durations arises from the same set of values. This notion results in the neglect of key vocal and sonic dimensions that, traditionally, are not notated. By considering the underwater singing practices of Snapper this chapter points the way toward those aspects of music that are inaccessible to standard notation but available to all of our perceiving senses. Snapper's work opens a window on the physical and sensory properties of singers' and listeners' bodies; on the spaces and materials in which sound disperses; and on these aspects' collective indispensability to singing and listening as lived experiences. Because sensory readings of singing and listening reach for dimensions of voice and sound that are difficult if not impossible to account for with conventional analytical methods, multisensory perspectives can enrich the analysis of musical sound in general, and vocal practices in particular.

Pushing the Limits of Voice and Body

Snapper's work experiments with (or perhaps against) the limits of her voice and body, challenging her physical abilities as well as her imagination.[2] The venues for her underwater operas range from bathtubs to Olympic-sized pools (see figure 1.1). The works range from solo pieces and duets to large-scale productions with choruses and dancers. *Aquaopera* was set for solo and duo performances, all of which invited audience participation; *Five Fathoms Deep My Father Lies* was modular in size, and ranged from solo performance with audience participation to performance with large-scale chorus and sound design elements; and *You Who Will Emerge from the Flood* was composed for soloist, full chorus, sound design, and keyboards.[3]

Snapper is a classically trained soprano, highly sought after by contemporary composers of complex music. Despite her mastery of vocal nuance and her success in the traditional music world, when it comes to her own vocal work, Snapper's main concern is the body and its mechanism and state; the sound is secondary. Snapper represents the third generation of vocal experimentation stemming from American classical music, part of a lineage of singer-composers that includes Laurie Anderson, Cathy Berberian, Meredith Monk, and Joan La Barbara (first generation); Shelly Hirsch, Diamanda Galás, Kristin Norderval, and Pamela Z (second generation); and Amy X, Gelsey Bell, and Kate Soper (third generation). Yet Snapper cites diverse influences such as comics like Carol Burnett, the punk vocalist Nina Hagen, the seventeenth-century opera singer and composer Barbara Strozzi, the composer and improviser George Lewis, and the artist Kathy Acker.[4] Additionally, in her nonsolo work—most recently, her involvement in the Human Microphone Project, part of the "Occupy Wall Street" protest movement—Snapper is intimately connected with the American composer and accordionist Pauline Oliveros, who, while not a singer, has composed a large body of work for group vocal experience. Oliveros's work, which seeks to erase the distinctions between performer and audience and between professional and amateur and to use "technology" (from the human body to instrument building and modifications and musical soft-

FIGURE 1.2 · Ron Athey on the Judas cradle, May 5, 2005 (photo by Manuel Vason).

ware and hardware systems) to break down those boundaries, opened a number of paths for Snapper.[5]

A classical singer who had trained for most of her life to gain complete control of her voice, Snapper began a journey toward unsettling, questioning, and challenging that foundation. By challenging sonorous traditions of opera, a highly formalized vocal genre that rests on assumptions of decades-long practice leading to high levels of control, Snapper also questions the utility of other areas of constraint. As I will discuss below, she questions the performance of gender and sexuality and the limitations of language in the face of nonnormative behavior. Her investigation aims to complicate her performing relationship with her instrument, her voice, by pulling the rug out from underneath herself, so to speak, and implementing techniques that would undo her hard-earned control.

Vocal Context and Influences

While these experimental practices seem to situate Snapper alongside composers who work with extended vocal techniques, Snapper understands her endeavor as a breakdown of technique rather than as its extension. She likens the process of breaking down her instrument to the instrumental preparation investigated by experimental composers of the late twentieth century, including John Cage, Oliveros, Annea Lockwood, and Cecil Taylor. To prepare a piano or guitar in this sense is to distort the instrument's capabilities by attaching alien objects to it, causing the instrument to create new and distinctive sounds. Similarly, Snapper distorts the sound of the operatic voice by penetrating, mutilating, or inhibiting the human body. For example, in *The Judas Cradle* performances, Snapper's vocal body is temporarily deformed by being tied upside down, while the anus of her collaborator, Ron Athey is penetrated by the Judas cradle as his soul is entered by the Holy Spirit (see figure 1.2); in *Five Fathoms Deep My Father Lies*, being underwater prevents Snapper from drawing breath. As a practice, preparation evidences both a curiosity and adventurousness about sound and a desire to interrupt and disturb human relationships with instruments and their histories. We might also imagine Snapper's vocal preparation as a way to remark on, negotiate, and play with the boundaries between nature and culture: between the female voice historically understood as uncontrollable or natural, and the operatic voice as refined and controlled.[6]

To offer a snapshot of Snapper's forerunners: La Barbara explores voice as an instrument; the celebrated Berberian, classically trained but often inspired by popular culture, investigates the voice's sonic range; Monk attempts to access the sonic space of the prelinguistic voice; and Galás scrutinizes the sounds of psychic space. Anderson is a composer with roots in a storytelling tradition. Her vocal work relies on techniques and effects enabled by microphones. In contrast to Galás, for example, who uses microphones to compositional ends, Anderson's vocal sonority is made possible by a close-miking technique. Galás also makes good use of the microphone and its ability to capture her voice, using a variety of microphones at once and processing each of their signals differently, but her basic vocal techniques do not seem to have resulted from microphone usage. In the same way, while Snapper makes use of microphones, her voice does not change when she sings with one (this is a fairly typical trait of classical singers).

Among Snapper's interlocutors are Z and Norderval. Z's work is completely dependent on microphones. Her trademark sonority comes from live sampling and playback supported by her interface system, the VocalSynth. With what appears to be a flick of her hand in the air, she layers voice and other sound samples. Z's work is also heavily based in storytelling. Her sound combines close-miked spoken voice with a sonority that resembles the classical vocal aesthetic (she employs a lifted soft palate, but not to the same extent as exclusively classical singers like Snapper). Norderval is a classically trained singer and composer who, like Z, uses custom-made technological systems to deploy vocal samples.[7] Both Z and Norderval make innovative use of technology to expand classical compositional and vocal techniques.

There are many more experimental singers of note, but those I have mentioned collectively form the orbit within which Snapper finds herself. Her work clearly emerges from this particular tradition of American vocal exploration—she even shared a teacher (the soprano Carol Plantamura) with Galás—yet her contribution to the contemporary vocal repertoire is distinct. While many of today's most innovative musical ventures, some of which are cited above, rely on digital technologies, Snapper has collaborated with the pioneering computer musician Miller Puckette, and her explorations engage corporeal, organic, and architectural technologies. Her work is concerned with the dynamic relationship between control and its loss—sonically, corporeally, and socially—and she investigates her material of choice through that preoccupation. Considered through the lens of performance art, in which Athey and many of Snapper's other collaborators participate (artists such as Paula Cronan, Elena Mann, Sean Griffin, Andrew Infanti, and Jeanine Oleson), Snapper's work may be heard in conversation with body art, specifically with art that modifies the body. Her work engages the body as both an instrument and a disruptive appropriation of culture. In the latter regard, the work resembles that of Marina Abramović, Vaginal Creme Davis, Karen Finley, and Annie Sprinkle.

Furthermore, beyond the obvious parallel of the marine environment, Snapper's work has much in common with that of other late twentieth-century composers who have explored the sonic possibilities of aquatics.[8] The composers whose work I believe Snapper's most resembles are not those who foreground the sounds of water, but rather those who work with sound in water.[9] Major composers who deal with the acoustic environment offered by water include Cage with Lou Harrison (*Double Music*, 1941), Max Neuhaus (*Whistle Music*, 1971), and Michel Redolfi (various works, 1981–present).[10] With Cage and Harrison, Snapper shares the notion of changing the sounds of a familiar source by immersing the sound source in water. With Neuhaus, she shares the desire to eject music from concert spaces and institutions and to showcase the sonorous possibilities of traditionally nonmusical environments. And with Redolfi, Snapper shares a fascination with adapting instruments, performers, and listeners to an aquatic medium. Nevertheless, to my knowledge, Snapper is the first to concentrate on singing underwater.[11] Of Redolfi's approximately two hundred underwater pieces, for which he has tried to perfect various custom-built instruments (mostly based on percussive principles), only one piece was created for a soprano. But whereas Redolfi is disturbed by the bubbles that escape from air-based instruments underwater, and therefore avoids such instruments altogether, for Snapper causing bubbles is part of the performative experience.[12] For her the idiosyncratic sounds of bursting bubbles, and the acoustic information those sounds offer about the bubbles' physical properties, form aspects of the music. Performing in an unfamiliar element forces the vocalist to confront the processes involved in singing on the most fundamental level: How do I get air? Do I emit the sound from my mouth or vibrate it through my bones into the water? How can I share the sound with my audience? Snapper addressed these questions through trial and error.

Prior to Snapper's underwater opera work, scholars (many of whom have influenced her) described a sensory complex that would accommodate additional registers of sound, voice, and the experience of listening. As we will see, this idea is often described as a difficult experience to capture in words or is expressed in terms of how it breaks with the normative way of understanding these musical moments. Roland Barthes insists that the voice has the capacity to operate outside the dependency of the semiotic sphere. With his landmark text, "The Grain of the Voice," he hints at sound's occupation of the tactile domain.[13] This entails a shift of emphasis from adjectives to verbs in the discussion of sound, as Christopher Small demonstrates.[14] Similarly, but arising from a different impetus, Suzanne Cusick mobilizes the notion of performativity in her suggestion that culture works its way "deep in the throat," and that certain vocal styles arise from the body's relationship to culture.[15] Thus, as Carolyn Abbate suggests, knowledge gained from hermeneutic analysis, while not completely divorced from the experience, can be completely contradicted by a given performance, or rendered irrelevant when a performer "offers up his body."[16] Steven Connor reminds us that the voice and ears are part of a multisensorial bodily landscape in which the transfer of experience from one sense to another (say, from hearing to touch) is natural and unavoidable—for example, one can even experience sound by biting on a vibrating rod.[17] That is, while Snapper's work is unusual in its clarity and heuristic, we see that scholars and artists never cease to grapple with the intersensorial aspects of sound.

Flood and Rapture

Snapper began her *Five Fathoms Opera Project*, "a series of modular, site-specific operatic performances," in response to an environmental disaster, which had been met with reactions ranging from apoplectic to indifferent.[18] Watching Hurricane Katrina on television from the West Coast of the United States, Snapper bore horrified witness to an emerging awareness of our changing climate, as fear of flooding and drought turned to a full-fledged politics of disaster. She watched Evangelical Christians absorb climate change into their idea of the rapture: the biblical end of time in the form of melting glaciers and rising sea levels. The Judeo-Christian perspective is predisposed toward a linear sense of time and the progressive inevitability of events. The end of the world is thus inexorable and often depicted as an uncontrollable flood—not as a gateway to cleansing and renewal, as with the flood of Noah's Ark, but as an eternal doom, an irreversible watery state. The element from which we ascended billions of years ago and that we depend on for survival, enjoy in recreation, and use as a means of transportation is also the unstoppable punishment that will obliterate humanity from the earth. Therefore, even as scientists search for clues about the beginnings of civilization, others predict the end of time, wondering: What are the signs? What deeds might trigger events of such magnitude? And how should we act when we are faced with the rapture? Snapper's practice questions the relationship between a progressive trajectory and the events that can be read as propelling it forward. For example, when a linear narrative is set in motion, she asks whether it is not a centrifugal force surrounding this narrative's trajectory that pulls events into it, to be read as its confirmation, rather than the events themselves causing the end time to draw nearer. Specifically, when homosexual practice is understood as causing a flood, the two are erroneously linked through a narrative, rather than causal factuality.[19]

Outraged by the uncommon, yet brazenly articulated, views of the supposedly inevitable suffering wrought by Katrina, and appalled at the inertia of the unaffected populace who—dry and warm in their living rooms—watched the flood unfolding, Snapper began to reflect on water's relationship with society, leading her to ask whether people had lost touch with water, its potential, and what it represents. Were they numbed by a media culture that profited from fear? Had they been pulled apart by each episode of paralyzing dread? "I think we need to take [opera] out of the opera house and bring bodies together. It can work against that separating damage," was Snapper's eventual response to these questions. She continued by describing her hope that opera, if ejected from the opera house and steeped in water, could infuse souls:

> The idea that water [always] represents emotions in some fundamental way is all over our language. The idea of being flooded with emotion, or storms of rage, or raining tears. It's very raw. [Water is a] technology that gets people feeling in a new way. My hope is to use that technology in a way that is more fresh and more immediate and really actually can work on people listening—which I think less and less happens in the opera house. . . . Maybe opera can help us to bind in new ways, to feel what we're feeling.[20]

Because water can represent extreme emotion, Snapper believes that to connect with water is to enhance our engagement with our feelings. She views underwater singing as a way to address a society distanced from itself and from emotion, paralyzed by the prospect of the end of time. Additionally, for Snapper, singing underwater is an adaptive strategy for basic and artistic postapocalyptic survival. She wonders if instead of accepting watery engulfment as the conclusion of our story, could one adapt to this new state? Snapper says: "I am interested in what it means to accept the end of things—instead of trying to keep things that are dear to us alive at any cost."[21] Thus, the *Five Fathoms Opera Project*, of which there are several versions, was born from the idea of adapting singing to the condition of the end of time and, through this adaptation, defying the end of time envisioned by a minority of Christian leaders.

Spilling the Truth

Several striking elements indicate a continuum between Snapper's *Judas Cradle* and *Five Fathoms Opera Project* (hereafter *Five Fathoms*). The unusual vocalizations that characterize both pieces arise from severe corporeal transformations. In *The Judas Cradle*, Athey's glossolalia spills out of him because God's presence has overtaken his body, and Snapper's trained voice breaks because she is hanging upside down.[22] In *Five Fathoms* Snapper's entire vocal repertoire and sonority are transformed by her aquatic immersion. Whether they are utilizing glossolalia in a queer masochistic performance, breaking the body to break through operatic training, or defying the end of time by learning how to sing under water, Snapper and Athey play with subverting regimes of body and mind not by escaping or averting them but by facing them to pervert them.

Both pieces are extreme responses to various manifestations of control exerted through terror. The Judas Cradle was a medieval European torture device, a pyramid-shaped apparatus onto which victims were lowered for penetration inflicted by their own body weight (see figure 1.2). *Five Fathoms* is also a response to horrific events. The title of the project, *Five Fathoms*, and one of the variations, additionally titled *You Who Will Emerge from the Flood*, combines an adaptation of a Shakespearean song—sung by the spirit Ariel to a shipwrecked prince in *The Tempest*—and a quote from Berthold Brecht's poem trio, *An die Nachgeborenen (To Those Born Later)* (1938).[23] Set to music by Hanns Eisler in 1984, Brecht's poem is addressed to survivors of world-annihilating tragedy, asking them to remember with understanding those who caused the tragedy. (Eisler's song is performed twice during *Five Fathoms*.)

Furthermore, the pieces subvert dominant narratives that significantly influence how the world appears to be configured. In *Five Fathoms* Snapper implies that even after the flood that Judeo-Christianity promised would destroy the world, humans can find the strength to survive and even to make music. In *The Judas Cradle*, as even the costumes indicate, she exaggerates conventional Western narratives of male and female sexuality to demonstrate how those narratives demand cruel and impossible confinements of the human body. Moreover, there are unavoidable parallels between the Judas Cradle and the similarly antique no-touch torture technique of waterboarding. The latter was frequently discussed in the early twenty-first century, as the U.S. government's antiterrorist program used waterboarding to obtain confessions. *Five Fathoms* is also a response to Judeo-Christian tales of the ultimate punishment—annihilation. The masochistic aspects of *The Judas Cradle* draw on a long history of Christian martyrdom, in which we find references to Jesus's hanging, pierced, and dismembered body.

Perhaps hoping to shed the undesirable trappings of her inherited Judeo-Christian culture, in *The Judas Cradle* and *Five Fathoms* Snapper forcibly divorces herself from the hyperbolic discipline of the operatic body—although it was not easy. She found that "the operatic instrument is actually incredibly tough" and difficult to disturb. "I started really delving into this idea of the prepared body with Ron [Athey]," says Snapper, recalling their collaborative 2005–2007 explorations that would culminate in *The Judas Cradle*: "We had a hell of a time trying to get my voice to break down under stress. We had me folding over jungle gym bars and contorting every which way before discovering that hanging upside down, with a slight arch to the back, will undo the vocal mechanism over the course of several minutes"[24] (see figure 1.3).

Through rigorous experimentation Snapper located the point at which she, as a singer, lost control, allowing her voice to take over as an autonomous, driven, and determined entity. Her own voice hastened her to places where her knowledge of singing and her artistic imagination could not take her. In other words, she discovered that allowing the physicality of her instrument, rather than prewritten instructions or preconceived ideas, to dictate the sound of her performance led her to new possibilities.

Opera, along with ballet, is arguably one of the most extreme arts, and it involves the "regimentation of the female body to attain an ideal."[25] By imposing additional challenges—such as hanging upside down or singing underwater—while still exhibiting mastery over the operatic idiom, Snapper questions the logic of the rules of bodily dominance set by operatic and balletic traditions and exposes their limits. Thus, by extension, she also questions the performance and practice of the female body, the female voice, and bodies within erotic acts and fantasies. And while in *The Judas Cradle* both performing and listening bodies are sonically, physically, and metaphorically penetrated through, in Amelia Jones's words, a "(masochistic) enactment of pain," the setting of *Five Fathoms* enacts the ultimate cradle, the embrace of oceanic depths, celebrated and feared since the dawn of history.[26]

Notably, in many Western myths the ocean is personified by the mutated female forms of sirens and mermaids.[27] Thus, in a connection to which I will return below, the flood—the apocalypse—is associated with the female in Western mythology. Snapper knows that her work will unavoidably be received within such contexts, so she deliberately alludes to the multiple penetrations of the female/oceanic body that enable it to absorb and exude sound.

FIGURE 1.3 · Juliana Snapper singing upside down in *The Judas Cradle*, May 5, 2005 (photo by Manuel Vason).

Situating Snapper's *Five Fathoms Deep Opera Project*

Despite Snapper's breakdown of her instrument in pursuit of new sonorities, at the core of her work is a committed, if fraught, relationship to operatic techniques, aesthetics, and institutions. Says Snapper: "It is an amazing feeling to sing operatically. All of this power gushing from your center! I trust it beyond any other means of communication. It is totalizing, erotic, uncanny. Every part of you is active, your insides are turned out. That's why I center my work within operatic singing . . . and explore the limits of the voice by directly addressing my body."[28]

To Snapper there is nothing more uncanny than operatic vocal technique, except possibly operatic form. In fact, the uncanny and fantastical are central to Snapper's practice: she metaphorically juxtaposes opera, arguably the most artificial of vocal forms, with dolphins' and whales' sonorous underwater communications, often used as examples of ancient forms of communication.[29] In this relationship, it is important for Snapper that she never becomes comfortable with singing in her newly chosen element, whether upside down or underwater. Not only the uncanny and fantastical, but also the states of discomfort and uncertainty of vocal outcome, suggest something extraordinary. It is an exciting but uncertain situation that keeps one at the edge and does not permit taking the environment and its rules for granted. To Snapper, the fantastical aspects of the operatic vocal sound and form, the newness of the elements she introduces, and the consequent uncertainty renders the operatic medium transformative: "I like struggling with a mastery that is no longer fully relevant (bel canto) and having to transform it into something else again and again. Singing underwater is still the most physically demanding because there are serious dangers attached to it—like bursting a lung or drowning [only three teaspoons of water in the lungs will do it]. Plus it just takes more energy to expel sound into water because it is so dense. But singing upside down was emotionally taxing. Maybe because your heart and head fill with blood, or for the way that failure happens gradually."[30]

Snapper sees herself as extending and dialoging with the operatic form, as part of its monodramatic lineage; she views her work as a continuation of the single-character tradition. In her scholarship she identifies Arnold Schoenberg's *Erwartung* (1909) as the predecessor of single-voice dramatic works, such as Francis Poulenc's *La Voix Humaine* (1958), Peter Maxwell Davies's *The Medium* (1981), and Luciano Berio and Cathy Berberian's *Visage* (1961).[31] In the interest of exploring media and sonorities, these monodramas were carried out as both live performances and taped pieces, in all of which the feminine body acts out madness operatically. Although these works most often feature women, the mad figure might also be a feminine man, or a man rendered effeminate by his madness—as exemplified in Davies's *8 Songs for a Mad King* and Julius Eastman's *Prelude to the Holy Presence of Joan d'Arc* (both 1981).[32]

Snapper calls this repertoire "hysterical" because of the extreme extent to which it "re-arranges the body of the singer" in ways that affect vocal quality.[33] This rearrangement "extends outward from [the performer's] body," affecting the other musicians' and the audience's empathetic bodies, "rearranging" our ingrained notions of music as an ineffable experience. Just as we may be lulled into believing that we are completely disconnected from the suffering that we watch from the comfort of our homes, so we can persuade ourselves that music, fleeting and seemingly intangible, has no lasting consequences. Snapper attempts to rearrange these beliefs through what she calls "hystericism," alluding not to an illness—like the hysteria historically assigned to women who did not align their behavior with prescribed gender roles—but to an approach to technique that deliberately harnesses physical responses to terror through music drama.[34] Snapper says she coined the term *hystericism* to describe "a non- or truly anti-discursive mode of vocal performance capable of transmitting things [that] symbolic systems (language, narrative, musical rhetoric) cannot."[35] While her performances are not about hystericism, her objective is to "harness the technology of hystericism to redirect the kinds of energy that propagate a growing culture of fear."[36]

Through hystericism, Snapper addresses how women are silenced, prevented from using their voices in ways that seem proper and natural to them, the ways this silencing plays out in emotionally lonely places, and her experiences as a woman with a fundamental distrust of language. After Snapper lost the ability to speak for a period of weeks at age nineteen, her relationship to verbal discourse and to social expectations grounded in language became deeply distrustful. While she could articulate words during this period, she could not form sentences or sing lyrics. Naturally this resulted in an inability to explain herself (and how can a nineteen-year-old explain that she has suddenly lost her grasp of language?). Though she attempted to communicate with her eyes and nonverbal sounds, she loathed the powerlessness that came with being half-mute. Snapper views her vocal compositions as reactions to her own abandonment by language and uncertainty in relation to the voice. Retreating from the everyday tool of language that temporarily abandoned her, Snapper's work restages situations of uncertainty, discomfort, and reaches for the uncanny and fantastical.

Singing Underwater

As she gradually adjusts to new self-imposed linguistic and physical constraints, Snapper's practice in the *Five Fathoms Deep Opera Project* involves continually pushing her body toward moments of surprise. She first experimented at home, in the bathtub; her first performances, too, were in tubs. "Once I got the hang of . . . well, I am still getting the hang of it," Snapper notes, "I started working with movement, different depths, different apertures" before moving into larger pools of water (see figures 1.4–1.6).[37] When Snapper described her process to me, we agreed that the best way for her to demonstrate being overwhelmed by a new environment was to take me through a comparable experience. So in 2010, I took a group of graduate students to the Standard Hotel in downtown Los Angeles. We gathered in the rooftop bar, one of Snapper's many performance venues, which featured fire-truck-red waterbeds and a large saltwater swimming pool.

Once we were in the water, Snapper took us through some participatory activities. The first had us form pairs; one person gently held the other underwater, while the person underwater made sounds (see figure 1.7). I was paired with Natalia Bieletto, who shouted—but with my ears above water I didn't hear her voice.[38] We tried another strategy: one person made sounds underwater while the rest of us put our heads and ears in, enabling us to hear him. We found that the deeper into the water we descended, the more difficult it was to sing high notes. Fast tempi were also difficult to maintain; Bieletto's attempt resulted in muddled sounds. Surprisingly, while sung sounds generally didn't seem very loud, small internal throat sounds were incredibly powerful. They boomed, beamed, and spread and were almost overbearing. These exercises demonstrate the extent to which the medium in which sound waves flow affects their characteristics: their speed, direction, and so on. It also shows that to register sound, the listening body, including the head, must be immersed in the material through which the sound flows.

The next exercise linked the six of us together by the arms: three participants stood in a line, with their backs against those of the other three. We sang in a drone-like manner, playing with our voices above the water, at its surface, and slowly descending into it. We felt the sonic vibrations largely through direct contact with each other's bodies. Of course sound also passed through the air and the water, but because the most immediate path was from one body to another: this was the sensation that overpowered us.

As we ended the day by gathering around the poolside fireplace, we discussed how taken we were with singing's different feel in a liquid environment.

FIGURES 1.4–1.5 · Above, top: Juliana Snapper singing in a bathtub (photo by Miles Rosesmire). Above: Juliana Snapper in *Five Fathoms Deep My Father Lies* in a tank at P.S.1 Contemporary Art Center/MoMA, New York City, March 15, 2008 (photo by Marina Ancona).

FIGURE 1.6 · Juliana Snapper in *You Who Will Emerge from the Flood* with João Tavares da Rocha and Hugo Veludo in an Olympic size pool, Porto, Portugal, October 10, 2009 (photo by Silvana Torrinha).

FIGURE 1.7 · The author (right) with Natalia Bieletto (left, under water), participating in *Aquaopera #5/Los Angeles*, April 28, 2010 (photo by Jillian Rogers).

Although some of us were singers with decades of training, we felt that little of our experience could effectively apply or even seem relevant underwater. We found that aural experience is predicated on our physical contact with sound waves through shared media—in this case water and air, flesh and bone. We noted that the shared medium makes a great deal of difference to how we experience the voice, and that the sound ultimately heard depends partly on what is sung, partly on the medium through which it passes, and partly on how our bodies interact with that medium. Connor's engagement with Michel Serres's work came to mind. "For Serres," Connor writes, "the body itself is caught up in a process of hearing, which implicates skin, bone, skull, feet and muscle. Just when we thought hearing was going to be put in its place, Serres evokes its own mingled or implicated nature."[39]

In other words, in Snapper's workshop we experienced what we already knew in theory. And although we had prior theoretical knowledge, we felt as though we had discovered that sound is a multisensory experience, tactile as well as aural. Snapper's exercises revealed that music making involves more than traditional theories and notation can capture, and even more than what current musical discourse can describe.

The Sensing Body in Relation to the Material World

Singing in water sounds so different from the way it sounds in air largely because, unlike electromagnetic waves of light, mechanical waves of sound require a medium through which to propagate. Consequently sound cannot travel through a vacuum.[40] Hence the speed at which sound waves travel depends on the density and compressibility of the medium through which they pass. Because higher densities and compressions engender slower speeds—and relative to air, water is very dense but nearly incompressible—the speed of sound in water is generally about four times faster than the speed of sound in air, with slight variations depending on several factors. These factors include water quality (distilled or salted, warm or cool) and hydrostatic pressure (which depends in turn on the distance below the surface at which an object sounds).[41] This is partly why Bieletto's up-tempo tune sounded muddled underwater, and why Snapper chooses slower tempi for underwater music than for the same music sung in air.

The specific relationship between our material bodies and the materials in which we immerse ourselves also affects how we experience sounds. In its unfamiliarity, listening underwater brings the relationship between sound, matter, and eardrum—which, in air, we take for granted—into relief. Because the density of human tissue is very similar to that of water, the eardrum does not provide the resistance necessary to translate underwater vibrations into tympanic movements—that is, into sound that eardrums can register. Thus, when we listen underwater, many vibrations pass through our eardrums without registering as sound. It is precisely because our skull bones are dense enough to convey the sound that those bones, rather than the eardrums, capture most of the sounds that humans do manage to register underwater. As a result the sound resonates in the body, going directly to the inner ear and circumventing the eardrum. Like air and water, the eardrum and skull bones are media through which sound passes, and by which its character is affected.

Bone-conduction theory explains that sound signals reach the inner ear not only via the eardrum's ossicle path, but also via the bone-conduction path.[42] Bone conduction can take place via air conduction, in underwater hearing, or through stimuli that set the body directly into vibration. While the early twentieth-century hypothesis that bone conduction is a crucial aspect of normative spatial hearing is no longer widely accepted, the German psychoacoustician Jens Blauert points out that when the sound component that reaches the inner ear is similar in strength to air-conducted sound, bone conduction does play a significant role. Examples of when sound can reach the inner ear with a strength similar to air-conducted sound include the use of ear protectors and the immersion of the body in a medium with a field impedance similar to its own—for example, water.[43] Favorable conditions for underwater hearing are enabled by the similarity in acoustic field impedance between water and the skull.[44]

The specific part of the body that registers sound also plays a role in its apparent directionality. For example, our ability to hear in stereo—two distinct signals, left and right—is the result of sound entering our bodies from two directions (through two ears). In contrast, when the inner ear registers sound via the skull bones, rather than with the left and right eardrums, the sound seems to be omnidirectional.[45] Because the sound waves vibrate the bones of the listener's body, her perception is that her own body has created the sound. The sound becomes a state or quality of the listener's body—in Stefan Helmreich's description, a "soundstate."[46] In effect, at an underwater performance where the audience and performers are immersed, the singer's body, the water, and the audiences' bodies connect through vibration to become one mass, a single pulsating speaker.[47]

The multiple iterations of *Five Fathoms* transmit sound along very different nodes. The paths of propagation depend on whether the audience is immersed in the water with Snapper or seated on bleachers by the pool. For example, in the 2011 Geneva performance of *You Who Will Emerge from the Flood*, when audience members sat on bleachers, they were not only removed from Snapper at a greater distance than was the case in the water-immersed mode, but also the number of transmissional nodes between Snapper and the audience increased, differing between sections of the piece. At various points Snapper's voice was emitted into air, water, a traditional microphone, and an underwater microphone. From these various nodes, her voice would be transduced (1) via air to eardrums; (2) via water to air—where most of the energy is lost and not transmitted—to eardrum; (3) via air to microphone to amplifier, speakers, air, and eardrums; and (4) from water to underwater microphone to amplifier, speaker, air, and eardrums.[48]

The contrasting propagations resulted in four distinct characters that engendered four distinct sounds and could potentially be understood as four different voices. While, needless to say, throughout these different transmissions the sound is sensed beyond traditional eardrum-based audition, it is the unusual situation of having different material afforded sensory experiences of the same musical segments that allows us to grasp that the identification of a musical piece is materially specific and dependent. Because so many discursive resources have been put into controlling and repressing the material, multisensory aspects of sound, we have stopped accounting for these aspects. But it seems that in limited cases, such as immersing ourselves in water with an opera singer, we can sense beyond the strong naturalization of our sensing of sound.

Musical practices that extend beyond normative perception tend to be marginalized as exceptional or as functional. But does it actually matter whether Snapper's work is framed as opera as opposed to, say, music therapy? That is, does Snapper's work reveal anything different from the revelations offered by practices that yield similar experiences, but that are framed as music therapy or alternative healing practices?

In summary, because musical discourse is in large part aesthetic, and only selectively engages with knowledge of sound gained in other fields, there is an unexpected twist to this story: the most important and revealing part of Snapper's work is not her need to develop a strategy for singing underwater, but that she sings underwater, deals in vibration, and still considers it opera. In that way knowledge about sound's propagation underwater and its particular material relationship to the body must be dealt with as it is articulated within a piece of music—as opposed to, for example, situations in music therapy, sound engineering, or medicine. Snapper performs in opposition to traditional conceptions of sound and music while working within traditions. Her insistence on defining her own creative collaborations as opera, and her legitimacy as an opera singer through her training and performance of more traditional repertoire, manage to keep what are considered extreme vocal articulations within the aesthetic. Thus, she opens up a space to think about aspects of sound and music that previously have been considered too liminal to have anything to say about music. Snapper's feat, then, is to offer extreme vocal articulations while managing to maintain her practice within a tradition-bound form of music such as opera. In so doing, she exposes the limitations of music analysis, an analytical framework that assumes that sound is propagated through air.

The Lived Body

In the midst of a crisis caused by the constraints of gendered and sexual life, during which the speech that supposedly reflects the vast range of human inner life was unable to communicate and describe it, we remember, Snapper lost her ability to use language. Her reaction and response to societal, cultural, and musical control involved devising incredibly difficult situations (singing upside down, singing underwater, and so on), challenging herself to master them within a demanding vocal tradition such as opera. By overcoming these impossible situations, she effectively breaks the narrative arches around gender, sexuality, and verbal communication that had trapped her. She upsets the conventional wisdom that dictates the forms of vocal communication by deciding where opera ought to take place. She directly questions both the idea of inevitability that a rigid, linear sense of time sets in motion and the tendency to slot events and even people into this notion of time and its progress. For me, a close reading of Snapper's work and practice has impelled an examination of those dimensions of sound that musicologists have not traditionally considered in earnest, such as materiality and its multisensory dimension. This examination in turn throws other inevitability narratives into question.

This type of analysis asks the subject to take leave of his or her absolute sovereignty and to acknowledge that he or she is both subject and object in the world, subject to material forces over which the mind and body do not have control. Following Shoshana Felman and Judith Butler, Michelle Duncan suggests that the "very 'scandal' revealed by psychoanalysis . . . propose[s] that the speech act undoes meaning, that it disrupts intentional knowledge through a bodily act," and that the nonlinguistic and nonreferential sound of the voice can move us to affect and action.[49] Wayne Koestenbaum shares how divas' vibrating voices worked on his consciousness via his bodily tissue in ways that were foundational to his budding awareness of his queer identity.[50] Accounts such as these, exploiting normative listening practices, are poignant examples of listening and thinking about listening beyond established paradigms, adapting Thomas Csordas's use of the term. That is, they describe instances of a new and "consistent methodological perspective that encourages re-analyses of existing data and suggests new questions for empirical research."[51]

"To be present in the world," writes Simone de Beauvoir, "implies strictly that there exists a body which is at once a material thing in the world and a point of view towards the world."[52] Beauvoir recognized that it is impossible to locate a body outside its performative representation of culture. In other words, she recognized that material in a natural state is a phantasm to which we do not have unmediated access. Rather, the materiality that we can access—which includes sound and the voice—is determined by ideas and representations that are unavoidably subject to power relations. The power relationship that we have seen play out in music analysis has pulled the eye and ear above the other senses, out of the full sensory experience that is listening to voice and sound.⁵³

In a response to feminist scholarship's debate about materiality, performativity, and nature in relation to the question of sex and gender, Toril Moi offers the image of the lived body (replacing categories of both sex and gender) as a means of illustrating the involvement of our bodily characteristics in the formation of a lived sense of ourselves.⁵⁴ This lived body is embedded in and subject to cultural forces at a foundational level. And it is this body, with a perceptual system tuned by a given culture, that is the perceiving conduit of sound. Quite strikingly, through her practice Snapper also addresses the possibility or impossibility of experiencing outside or beyond the situated sensorium. If she does not fully present the body with a new sensory complex by exposing it to situations such as singing underwater, which transcend the dominant narratives and rules, at the very least she questions whether the current narrative is entirely waterproof.

By descending into water with excessive performative expressions, costumes, set designs, and vocal sounds, and by inviting her audience to accompany her, Snapper confronts the pervasive cross-cultural ambivalence about the female body head-on. She challenges notions of this body's dangerous ambiguity, notions that have survived across geographies and cultures for millennia in stories of sirens' and mermaids' seductive, enveloping voices. Moreover, the idea of the underwater female body and its material form symbolize, literally and figuratively, the feminine embrace and the allure of all humans' complete dependency on a woman's body while in utero.⁵⁵ Snapper's performance points to inconsistencies in the stories we are told about how the world works, and it defies the rules surrounding vocal performance and operatic practice. From her work we have learned that (1) sound does not exist in a vacuum but is materially dependent. Therefore, (2) the transmitting medium (for example, water versus air) and the combination of different materialities (such as the body in relation to water versus air) affect the sound's propagation and hence its actualization. As a consequence, (3) listening is materially dependent. And, moreover, (4) we can arrive at these conclusions about sounds and music only if we investigate them in a material and multisensory register. The dominant discourse about sound, the assumptions on which traditional analysis is based, and my proposed material and multisensory notion of sound summarize the comparison between the perspectives of the figure of sound and the practice of vibration in the introduction.

As Susan McClary observes, "our music theories and notational systems do everything possible to mask those dimensions of music that are related to physical experience and focus instead on the orderly, the rational, the cerebral. ... The fact that the majority of listeners engage with music for more immediate purposes is frowned upon by our institutions."[56] While much has changed (in part as a result of the impact of her own research) during the twenty-plus years since McClary penned this insight, work effectively and affectively addressing these dimensions—as well as useful, efficient, and compelling tools with which to examine them—is still needed.

The above scholars relate experiences that exceed the communicative capabilities of current linguistic and analytical schemes. For example, the "scandal" referred to by Duncan is, I believe, a grasping toward words and frameworks through and from which we may attempt to understand such experiences. The sometimes incomplete quality of these scholars' works is a symptom of the vastness of the project on which they have collectively embarked. Yet their work accomplishes a crucial task: it points toward vocal aspects to which current ontologies, based in sound and/or vision alone, cannot do justice. It is by grasping at the ungraspable that we begin to crack seasoned, hardened paradigms. While this moment is often illustrated by an experience that is difficult to capture in words, or captured only through describing the ways in which it breaks with the normative understanding of these musical moments, these experiences are only glimpses. Snapper's underwater vocal practice shatters the premise on which music analysis and quotidian notions of our daily experience of sound rest: that of a stable, knowable sound.

Snapper's expanded notion of sound presents a new perspective akin to a paradigm shift. I envision the first stage of this process of breaking into a new ontological perspective as akin to a Heideggerian hermeneutic circle, which gradually moves us toward a multisensory understanding of sound. Traveling around the circle, we alternate between two different perspectives: one finding an experience (often a displaced experience, like Snapper's underwater opera) inexplicable within the current paradigm and another questioning the paradigm that fails to encapsulate the new experience. That is, we use this impossible-to-account-for experience to pry open the paradigm that excludes the experience. This circular and repetitive movement from particular experience to ontological questioning and back again is analogous to the hermeneutic circle.

Additionally, following Thomas Kuhn, I argue that the gradual paradigm shift that occurs through this quasicircular alteration of perspective resembles movement from one scientific paradigm to another.[57] Kuhn observes that a scientific revolution takes place when scientists encounter anomalies unexplainable by the universally accepted paradigm under which advances have thus far been made. A paradigm shift, then, is a matter not just of a new theory taking hold, but also of the emergence of an entirely new worldview from which new theories and perspectives flow. While anomalies can exist in any given paradigm without fundamentally challenging it, here we see an accumulation of significant anomalies sufficient to cause the discipline to be thrown into a "state of crisis," to draw on Kuhn's terminology again.[58]

The state of crisis identified by Kuhn as the launching pad for a new paradigm refers, in our case, to the dead end in sonic research and discourse. While some might argue that this state of affairs is not sufficiently significant to constitute a paradigm shift, I believe that much scholarship on sound, music, and voice has descended into a state of crisis through continuously opposing vision and audition.[59] Sterne has succinctly summarized the binary distinction that still results in discursive battles between eye and ear. In a return to Enlightenment ideals, "the audiovisual litany," as Sterne puts it, "idealizes hearing (and by extension, speech) as manifesting a kind of pure interiority."[60] For a number of years, one principal source of fuel for sound scholars has been the question of whether there might be a way to look beyond the visual and aural binary when thinking about and through sound. While the intense tendency to back away from visually based paradigms seems to have diminished, we are still in need of useful models with which to think through sound. As I suggested above, a new paradigm, or simply a useful and usable framework from which to think through the multitudinous experience of sound, can only be created if we think about sound from an altogether different perspective: rather than conceiving of voice and sound as phenomena with fixable identities, captured and held by the eye or ear, instead we must understand that we are party to, and partake in, a process and an experience.

What I seek to offer in the remaining pages of this chapter is a general analytical picture drawn from these experiences. Specifically, I apply the notion of sound suggested by Beauvoir, Moi, and Snapper—namely, that sound is sensed by the material, lived body—to a classic maritime tale of listening and voices. I ask if, as a multisensory perspective suggests, we are all ears, will Odysseus's story, hinging on eardrum-based audile techniques, hold up?

We recall Homer's Odysseus binding himself to the mast of his ship, having instructed his crew not to untie him even if he begged, while the crew's ears were filled with wax so that they would not be fatally tempted by the sirens' alluring voices. This story ostensibly demonstrates Odysseus's bravery and cunning by showing him outwitting the sirens while also enjoying their voices.

Music's Material Dependency · 51

Some have already read this story against the grain. Let us consider three especially relevant readings and, finally, conduct a brief reading within the new paradigm.

Reexamining the Encounter between Odysseus and the Sirens

Ruminating on the reasonable idea that no amount of wax could keep out the sirens' tempestuous voices, Franz Kafka suspects that there was actually nothing to be heard by either the sailors or Odysseus. Kafka believes that the sight of Odysseus's triumphant expression as he sailed into danger with his genius plan in place rendered the sirens silent—or perhaps they decided not to waste their voices on him. This view suggests that Odysseus in fact heard nothing; instead his desperate gesticulations to the sailors, ordering them to untie him because of the pull of the sirens' voices, enacted one of the tallest tales a man ever told. In Kafka's words, "the Sirens have a still more fatal weapon than their song, namely their silence"—and Odysseus "did not hear their silence."[61]

So here we are presented with two different accounts of how Odysseus and his men were able to travel close enough to expose themselves to the sirens' voices,[62] yet escape their fatal allure. The difference between the stories turns on two points: how the men were kept safe from the tempting voices and what Odysseus heard. Homer's account suggests that, because of his cunning, Odysseus is the only person who ever lived to tell of the sirens' incredible voices. Kafka claims the whole thing was a setup that has conned readers and audiences for thousands of years. The men sailed safely by because the sirens did not actually sing.[63] (But, having gone to the trouble of being tied to the mast, Odysseus would have been the last to admit that his efforts were in vain.) Hence, in both Homer's and Kafka's accounts, the sailors were prevented from bearing true (auditory) witness to events. Moreover, both interpretations assume that hearing takes place, and that Odysseus hears exactly what he is presented with—even if it is silence.

For both Mladen Dolar and David Copenhafer silence is the opposite of speech, singing, and sound. Both scholars assume that the absence of one is the presence of the other, and also that the existence of one always already offers the potential of the other. For Dolar, reading Kafka, the voice at its purest exists in silence.[64] For Copenhafer, Kafka's casting of Odysseus' wax-filled ears pretending to resist nonsounded sound constitutes a situation in which it "becomes necessary to play with one's potential to listen and to not listen."[65] While both readings and musings on Kafka's version of Odysseus's encounter are intriguing, to me the notion of a default to silence as the opposite of the potential impact of voice lacks imagination in terms of the human voice's sensory range. It also strikes me as too simplistic that a slightly mysterious element of this story, the reason assumed for Odysseus's triumph, is silence. Is Odysseus—or, for that matter, are we—up to that task? It is significant that the element on which this story turns, the cause for Odysseus's triumph, is silence. "The politics of silence often assumes a conservative guise and promotes itself as quasi-spiritual and nostalgic for a return to the natural," Steve Goodman writes. "As such, it is often Orientalized and romanticized tranquility unviolated by the machinations of technology which have militarized the sonic and polluted the rural soundscape with noise, polluted art with sonification, polluted the city with industry, polluted thought with distraction, polluted attention with marketing, deafens teenagers and so on."[66] Goodman's list contains well-rehearsed pure sites that are liable to contamination. Reading Odysseus, it is worthwhile noting that the nonsilence—that is, the nontranquility and the destructive force—is, in Odysseus's case, the sound of feminine gendered voices. Therefore, to that list add "polluting male rationality with the female voice."

Let us return to Kafka's suspicion about whether the sirens made any vocal sounds by recalling Max Horkheimer and Theodor Adorno's reading of Odysseus's exploits.[67] These scholars indict Odysseus for initiating a dynamic realized much later by capitalism, accusing him of finding no more than entertainment in the sirens' song, a protobourgeois reduction of the song's unique enchantment to the "longing of the passerby" (*Sehnsucht desser, der vorüberfärt*).[68] Drawing on Adorno and Horkheimer in writing about Debussy's "Sirènes," Lawrence Kramer suggests that an Odyssean dynamic exists between the modern concert audience and the music it consumes: "Fixed in his seat at the concert hall, the listener becomes the modern form of Odysseus tied to his mast, for whom the enchainment of the body makes possible the enchantment of the mind."[69] These readings, as Adriana Cavarero has pointed out, are anachronistic in their consideration of the "bourgeois."[70] But a view of Odysseus bound to the mast as a restriction of perspective wherein voice becomes commodity is nonetheless a perceptive observation about our general tendency to confine the experience of voice to a single physical sense.

The new paradigm I propose would ask us to examine how listening functions in these stories from a multisensory perspective.[71] If hearing is strictly confined to the eardrum, either Homer's or Kafka's version may be accepted without hesitation. Nevertheless, within the paradigm of voice and sound that I suggest—which refuses to assign stable, measurable identities to sonic phenomena and refuses to associate each physical sense with only one region of the body — both Kafka's and Homer's plotlines collapse. Applying this new paradigm, we could ask: Could not the sailors, despite the wax in their ears, sense and be affected by the vibrations of the sirens' voices? "The song of the Sirens could pierce through everything," says Kafka — but volume doesn't seem to be the pivotal point here.[72] Research on prenatal hearing has confirmed that reactive listening takes place as early as sixteen weeks into fetal development — in other words, eight weeks before the complete formation of the ear (which occurs at twenty-four weeks). This has led researchers to believe that hearing begins in the skin and skeletal network. Listening begins with the rest of the body in a "primal listening system" that is only later "amplified with vestibular and cochlear information."[73] In other words, our sense of hearing relies not only on excited eardrums, but also on sound conduction and vibration throughout the body.

I would respond thus to Homer's story: If the sirens indeed sang, the sailors would not have been protected by their wax-filled ears, as their entire bodies would still have conducted the sound. If Kafka is correct in his suspicion that the sirens did not sing, the most potent question is not about Odysseus's honesty but about his hearing: Did Odysseus know that they were silent, or did he honestly believe that he heard them? From this we can extrapolate the question of the aesthetic object: What do we have the capacity to hear when we are enmeshed within a certain sonic paradigm? Can we hear and sense what is right in front of us, beyond the given paradigm? Perhaps the sirens were silent, but Odysseus was trapped by the idea that they would sing; his belief prevented him from hearing their silence.

Are We Still Odysseus?

On my theoretical table, I have laid materials spanning thousands of years — Snapper's underwater opera and Odysseus's challenge — as vehicles for comparing the implications of traditional and multisensory vocal paradigms. Snapper's challenge of mastery through a strategy of technical proficiency, the success of which disproved traditional rules, prompted my shift in analytical focus from notions of stable sound and implied notions of a priori knowable works to multisensory aspects. According to this new analytical framework, Odysseus was out of touch with the reality of the situation. Odysseus's version of the story suggests that because he was so smart and had planned everything so well, he was able to trick the sirens. In general, he implies, it was his abilities (rather than luck) that led him to overcome the multiple challenges presented by his journey. But analyzing Odysseus's encounter with the sirens from a multisensory perspective calls his side of the story into question and begs us to probe his motive. Looking closer, we may see that Odysseus imagined these manly challenges and his triumphs over them. That is, when he felt that his life was in danger as he attempted to break away from the mast to which he was tied, the perceived danger was a phantasm, existing only in his head; the sirens did not waste their voices on him. But this illusion was not only of his making. From a multisensory perspective, his error did not lie in the belief that he had conquered the sirens; rather, it lay in believing the story of the sirens' voices that had been passed down to him, in allowing this tale to go unquestioned, and in profiting from such a narrative. Rather than perceiving the material specificities of the particular circumstances in which he found himself, Odysseus heard what he expected to hear.

In contrast, when singing under water, Snapper set out to question master narratives regarding gendered expressive possibilities, how to deal with the notion of the end times, and how and where to practice voice and opera. By mastering situations that were supposedly impossible to master within operatic practice, Snapper felt freed to create a logic and a worldview that worked for her. Applying the perspective derived from analyzing her music to music scholarship, we may see that Snapper's work contributes to dismantling common analytical tools and begs for new ones. When we undertake the perspectival shift that Snapper offers, we realize that the being of sound can no longer be assumed to be stable throughout time and circumstance. By extension, we see that we cannot fully account for the creation and functioning of the world according to master narratives.

Snapper's underwater opera work shows us that air has been naturalized as a material transducer for sound, and therefore suggests that sounds are contingent on the material circumstances in which they are created and experienced. By highlighting the material aspects of sound and their reception, Snapper reminds us that what we hear depends as much on our materiality, physicality, and cultural and social histories as it does on so-called objective measurements (decibel level, soundwave count, or score)—which are themselves mere images, icons, and metaphors. Indeed, the experience of sound is a triangulation of events wherein physical impulses (sonic vibrations), our bodies' encultured capacity to receive these impulses, and how we have been taught to understand them are in constant play and subject to negotiation. Experience cannot support a stable ontological explanation of sound or music; rather, as sound cannot exist in vacuum (as alluded to in the chapter's epigraph) each such account is a composite manifestation of our understanding of sound at a given material moment in time and place.

On the one hand, each aspect of sound on which Snapper invites us to reflect is a well-known sonic phenomenon. These underwater operas remind us that music affects us in many ways beyond those for which current analytical schemes can account. On the other hand, despite the seeming extremity of Snapper's underwater practice, it is particularly illustrative of the inability of sound, music, and voice to be autonomous, because they are always already materially dependent. Moreover, Snapper's is a self-conscious music practice that insists on the operatic normative label. Pulling singing, listening, voice, sound, and music into a multisensory schema addresses the crisis of audiovisual centrism and the dependency of each sense on the other and, simultaneously, the binary between thinking about music as for the eye or the ear. Indeed, by letting go of the idea that a 440 hertz A, lasting a quarter-note in the metronome count of sixty, is always precisely *this*, or remains precisely *this* regardless of who is listening, where, and under what material conditions, we relinquish the notion of a stable analytical basis and an even more fundamental certainty.[74] Because the traditional paradigm denaturalizes sound's material specificity, it assumes the ability to know a priori. Snapper's underwater vocal practice shows us that what forms the basis of analysis is not only a set of measurable musical parameters defined prior to experience; instead it is an unfolding process bound to a materially specific, contextually grounded and formed sensory complex.

Analytically, then, does it matter whether we know that sound is multisensory? Or, for that matter, whether we indeed sense the sound beyond aurality—a state to which this book's title, *Sensing Sound*, alludes? As we will see in the remainder of the book, I believe that the multisensory aspect of music is powerful, working on us whether we are aware of it or not. It is precisely because of the privileged position given to the aural mode of sound within musical contexts that other modes such as vibration are in danger of being conceptualized as extramusical, or as extreme and liminal. In this view, a mode such as vibration is understood as nonintegral to explaining musical phenomena, with the result that vibrations often fail to be accounted for as part of musical or aesthetic experience. Instead, vibration is commonly understood as a science of sound—a liminal, nonaesthetic sonic phenomenon (largely relevant to, say, heavy bass music or deaf community music appreciation)—or as offering a subpar aesthetic contribution (for example, music therapy).

The straightforward answer to the question of whether it matters that we know that sound is multisensory is that, in terms of sound's effect, it does not matter. Yet it does matter when sound is not recognized as multisensory, and its presence is thus addressed in other terms—for example, meaning, value, an impossible-to-analyze feeling, and so on, or the aspect of music that McClary has described as "listeners engag[ing] with music for more immediate purposes."[75] In other words, by not only privileging but even naturalizing the aural mode of sound in music analysis, discourse, and definition, we have also aestheticized sound in a total and limiting way that, in the end, will stand between us and the possibility of gaining full understanding of the rich phenomenon of music. Investigating the multisensory aspects of musical practice and experience is one productive path toward more fully understanding our meeting with and participation in what we call sound and music. Considering multisensory aspects of music will yield deeper insights into "sound and music cultures in relationship to power," which Goodman has dubbed the politics of frequency, and which I would slightly rephrase as the politics of aestheticizing the sonic mode of frequency at the cost of failing to include vibration in normative musical experience.

In summary, denaturalizing air as the default medium for the material transmission of sound through examining Snapper's vocal practice has allowed us to take the first of three steps, which are distributed over the first three chapters of this book. These moves clarify (1) the material specificity of sound, (2) the spatial-relational acoustic dimension of sound, and (3) the merely symptomatic role of sound. We see that by taking seriously sound's material dimension—not only in the scientific realm, but also in the context of aesthetics, voice, and music—we may understand more about voice's, and music's, impact. This will eventually lead us to the concluding chapter, which brings these notions to bear on our capacity to understand ourselves in relation to the world and other human beings.

2

THE ACOUSTIC MEDIATION
OF VOICE, SELF, AND OTHERS

One. I am seated in the Walt Disney Concert Hall, in Los Angeles. On the stage, Meredith Monk and Vocal Ensemble move around, singing lines. The last phrases are sung as the performers slowly lie down, flat, on the floor. It looks out of place—nothing more than several people deciding to lie down on the Disney Hall stage. Nothing in my previous concert-going experience has prepared me for how to approach or interpret this. I feel uncomfortable. The vocal lines sound simple—that is, undeveloped. I feel as unprepared to make sense of these sounds as I do to watch the unfolding scene. I wonder why I have this profound feeling of inability to deal with this event. Having experienced *Songs of Ascension* five times in two previous locations, why is the experience of the piece in this location so radically different?

Two. I am at Union Station, again in Los Angeles. Through headphones I hear Christopher Cerrone's *Invisible Cities*, featuring an eleven-member orchestra and up to eight voices. While the music is performed live in the station, the audience never inhabits, in person, an acoustic space in which all the voices and instruments sound at once. When I am close to a singer who is singing, I feel once removed from the performer, as I hear his or her voice with more strength and presence from the headphone signal than from the acoustic transmission. As I allow the carefully curated acoustics in the headphones to pull me into the piece's sound-designed world, I feel distanced from the site.

Why does a site-specific piece makes me feel more disconnected from the live singers than any other live performance I have experienced in the past?

Like chapter 1, this chapter deals with musical experiences that I wanted, at first, to dismiss on an aesthetic basis, yet the conundrums they offered lay beyond aesthetic preference. By trying to understand why what we recognize as the same piece of music could have such different effects in different halls, and by investigating the gap between the sense of presence in an acoustic performance and a microphone-and-headphone-mediated rendition of a piece, I learned more about what constitutes the figure of sound: I finally understood that acoustics offers more to us than delivering optimal sound and optimizing sound. I learned that acoustic and spatial specificity also take part in giving form to the figure of sound. That is, the figure of sound is made up not only of naturalized notions about pitch relations and a limited set of behaviors in limited material conditions (air). Our notion of the figure of sound is also bound up with a naturalized acoustic identity (including parameters such as reverberation and clarity, which I will discuss further below), location, and distance between the sound source and the listeners.

Acoustic mediation of sound and habituations to it are not limited to informing and shaping our sense of music. Rather, they profoundly mediate our experience of self and others. Acoustic mediation of sound profoundly influences the ways we assign meaning to self and others, and the ways in which we conceive of our own and others' positions and relationships. I wish to offer tools with which to analyze the acoustic mediation of self and others—hence contributing to understanding how the politics of difference is structured.[1] By considering the listening choices the aforementioned two operas offer, this chapter provides the second stepping-stone toward the book's final questions: How are ontologies and epistemologies of voices acoustically mediated? How does the acoustic rendering of voices play into formations of subjectivity and intersubjectivity? Moreover, how does that mediation influence, limit, and invite certain experiences of other and self? To consider these questions, we will first consider the ways in which the acoustic is normalized into the figure of sound.

In this chapter, then, I explicate the ways in which acoustics has been standardized in public concert and opera halls—to which the way notes were standardized according to the tempered scale might offer a loose parallel. I discuss how the figure of sound, including its expected aspects of space and relationships in spatial terms, is experientially reified through the building of spaces and the development of other acoustic determinacies, intellectualized through concepts and vocabulary, and reconfirmed and ossified through experiences guided by these ideas. While music—and sound, more generally—has always been experienced in a variety of spatial-acoustic configurations, because of the privileged status of the repertoire played in the symphony hall and the elevated status of the concert-hall listening experience, it is the kinds of sound and music that are played in that acoustic condition that formed the basis for the listening, discourse, vocabulary, and concepts that we use to make sense of music today. I show that by eschewing such notions, the aforementioned productions of *Songs of Ascension* and *Invisible Cities* offer audience members a choice of listening and relational stances. It is the existence of that choice that I wish to point to in this chapter.

"Berlin Stinks"

The history of concert-hall acoustics in itself is beyond the scope of this book.[2] However, drawing on robust research from musicology, architecture, and architectural acoustics, I offer a few key moments. I chose these moments to exemplify how nonmusical, nonsonorous dimensions were key to the constraints placed on the acoustic conditions of public concert venues; the formation of, and consequent commitment to, an acoustic sensibility; and the formalization of today's acoustic norm. I also chose these moments to exemplify the construction that led us toward a unified Western understanding of good acoustics. While I mainly discuss halls and acoustic conditions, it is important to bear in mind that these experiential repetitions of music that is sounded—the sound's specific acoustic conditions—are inseparable from the ways in which people and critics heard the music, and that the acoustic condition is inseparable from what is otherwise experienced, articulated, and conceptualized regarding the sound. Moreover, as sound is heard, impressions articulated, and concepts formed, these concepts themselves direct further impressions of music and limit our thinking about others.

The acoustic dimension of the figure of sound can begin simply as a practical question involving the optimal acoustic for a particular repertoire. Then, through repetition, the experience and standard concepts used to describe those acoustic phenomena and experiences burrow into our perceptual repertoire, and further language is formed around these experiences. In turn, these linguistic, conceptual, and perceptual frames inform expectations and further experiences—in short, they lay the groundwork for the acoustic dimension of the figure of sound. (I will return to the specifics of the conventions related to the figure of sound's acoustic dimension, a concept that will be developed throughout this chapter.)

By the spatial-relational and acoustic dimension of the figure of sound, I mean simply that the framework within which we imagine sound, and that we subsequently fit around the sounds to which we are exposed in daily life, is not limited to pitch and its duration (and/or to rhythm and meter). Included in our practices of sound is an acoustic dimension—which may be simply described as the length of the reverb and the sense of clarity (which I will discuss at much greater length below). Unlike pitch and duration, however, the spatial-relational and acoustic dimensions are noticed and called out only when they are nonnormative. That is, when a sound is too close, dry, wet, or uneven or exposes an unusual nonnormative feature—for example, a whispering arch— we become conscious of it and can overlay it with a particular meaning. Again, when a sound adheres to the normative spatial-relational and acoustic aspect of the figure of sound, we do not notice it. What is the process by which select sounds become naturalized?

The spatial-relational and acoustic dimensions of sound are naturalized within distinct sonic, performative, and listening practices. The music I will discuss in this chapter, and indeed in the entire book, is heard and conceptualized within the framework of Western classical music. As such, the spatial-relational and acoustic dimensions of this music's figure of sound have been formed partly through public concert practices. Historically, concert music was performed outdoors or in existing enclosed venues such as churches, theaters, or palace rooms.[3] The York Buildings, the first public hall with the explicit purpose of housing performances of music, was erected in 1678 London.[4] While this hall could seat an audience of two hundred, the later Hanover Square Rooms, where twelve concerts featuring Haydn in performance were presented in 1791, accommodated 900. As the orchestra grew in size, and the concert was transformed into a public format, demand arose for additional halls sized to fit increased audience capacity.

The first dedicated concert halls were constructed in the eighteenth century in Oxford, London, Hanover, and Leipzig. The sheer size necessary to hold the growing audiences and the desire to also offer a view of the orchestra led to seating plans and overall shapes that created "uneven acoustic results." (With its horseshoe shape, the Royal Albert Hall in London is often mentioned as a famous example of less than optimal acoustics.)[5] At the beginning of the eighteenth century, with the spread of concert music as entertainment throughout the continent, concert halls were built in Berlin, Vienna, Stockholm, and elsewhere.[6] And while, to a large extent, the acoustics of the halls were the main concern, additional criteria such as sight lines from the audience to the stage and between certain segments of the audience, linguistic clarity, and audience capacity were also important in defining the concert hall's overall configuration.

Therefore, while the criteria that shaped the building of the halls — including shape and size, sight lines, and seating arrangements — were not all acoustic concerns per se, the acoustic criteria were adjusted to also support other criteria. For example, explicit public concert utility concerns constituted a huge break from previous indoor settings of music. In other words, while previous indoor concert settings were explicitly exclusive, the public concert setting was explicitly inclusive and hence required a larger number of seats.[7] And while moving from privately organized to public events, viewing of the stage and seating arrangements was not detached from power, since relations and social distinctions were still performed through the construction and patronage of these buildings. Therefore, while not limited to royalty and their private guests, the new concerts featured many of the same or similar seating arrangements and concerns as earlier private functions had.

To meet demands regarding audience capacity, sight line to the performers and select public figures, and acoustic conditions, large concert halls were increasingly modeled on the parallelepipedic form of drawing rooms and Protestant churches.[8] So common was this form that it became the first referenced acoustic model, known as the shoebox. Within the shoebox, the orchestra is placed in one of the short ends of the box, facing inward, and the audience is placed directly in front of the orchestra, facing it. In other words, a triangle of (1) concern for audience capacity, (2) audiences mainly seated in front of the orchestra, and (3) acoustic concerns drive the general design of the hall.

In addition, concerns arose regarding the overall sonority of the music, giving rise to linguistic criteria. The 1778 horseshoe design of Milan's La Scala was copied throughout major cities in Europe and beyond, because the "relatively short reverberation time (1.2 seconds) allowed composers a unique creative privilege: to write opera with one kind of acoustics in mind."[9] Operatic repertoire's acoustic requirements were distinguished from orchestral requirements. To preserve the intelligibility of the libretto, sung at sometimes extremely high musical speed, the reverberation time needed to be relatively short. This would allow for the free sounding of one syllable without allowing the previous syllables' reverberation to mask it.[10] This design gave us examples such as the Naples San Carlo, Paris Garnier, London Royal Opera, Vienna Staatsoper, Munich Staatsoper, and Philadelphia Academy of Music. In these houses, the acoustics is guided by the need for audiences to hear the works in their native tongue. In contrast, where audiences traditionally hear operas in the original setting and with subtitles, the need to hear each syllable unhampered is less important, and, consequently, the reverberation can be more "attuned to the music than to following the libretto."[11] Once these halls were created and housed performers, those performers, the composers, and audiences adopted the sound of the halls as the norm that defined good sound. This led to a self-perpetuating cycle.

Composers write music for the kinds of acoustic spaces they know. As a result, the kind of sound that is considered good is reinforced, and the kind of sound that is considered natural is conventionalized. As Thurston Dart notes, "even a superficial study shows that early composers were very aware of the effect on their music of the surroundings in which it was to be performed, and that they deliberately shaped their music accordingly." He goes on to divide musical acoustics into "resonant," "room," and "outdoor" and offers, as examples, plainsong along with the harmonic styles of Léonin and Pérotin—pointing to the latter's style, even more specifically, as shaped for the exact acoustic space of the cathedral of Notre Dame in Paris. In contrast, *ars nova*'s intricate rhythms and harmonies were composed with less resonant acoustics in mind. Dart offers brass consorts by Matthew Locke as examples of music written for outdoor acoustics, pointing out that a composer's style is distinguished by the acoustic for which the music is intended.[12]

Natural sound, then, is the hallmark of a naturalized spatial-relational acoustic. We hear reports that musicians develop intimate familiarity and relationships with a particular acoustic. That is, when music is written for a particular acoustic, and the acoustic condition in which a musician plays conforms to the dominant acoustic paradigm, expectations of what is a good or bad—or normal or abnormal—acoustic are reinforced, creating a limited range of acceptance for acoustic divergence. For instance, as Robert Fink has noted, we can witness strong values around so-called natural acoustics, expressed in a rather explicit manner, in the discussions about Los Angeles's Walt Disney Concert Hall. In broad strokes, the debate focused on the seemingly irreconcilable needs for an urban development project with an innovatively shaped hall and democratic seating that would not divide patrons into obvious patterns based on price tags. Because—for musicians such as Sidney Weiss, the Los Angeles Philharmonic's concertmaster—the hall was "an instrument, like his violin, and its primary function was to enable the musicians on stage to play beautifully," familiar acoustic condition was imperative.[13] In response to inquiries regarding new music's possible acoustic needs, the conductor Zubin Mehta explicitly discounted the possibility, "appealing several times to a repertoire-independent ideal of 'natural' sound which he called a 'normal good room acoustic' or a 'good-sounding hall.'" Summarizing the year-long and passionate discussion, Fink notes that putting "30 years of playing experience behind 'the classical hall, the rectangular hall,' . . . Weiss, like 90% of his colleagues, just wanted some kind of shoebox to play in."[14] In other words, behind the desire for the shoebox, training in and the naturalization of acoustics are at work. Additionally, in Mehta's comment we can discern the sentiment that sound is just sound ("pure sound") and, by extension, music or compositions that fail to work well with the dominant acoustics constitute bad music rather than being mismatched with the acoustic condition.[15] The problem was ultimately solved through, to paraphrase Fink's description, a box with a wrapper.

Evidence of the strong and established sensory complex built, and values established, around acoustics has been poignantly summarized by Clifford Siskin and William Warner, who noted: "Enlightenment is an event in the history of mediation."[16] Acoustically, modern and contemporary sound has been increasingly unified around the two-second reverberation. Weiss's and Mehta's very specific preferences and concerns about the acoustics created through the process of limited acoustic mediation of music exemplify the formation of and membership in an acoustic community in which audiences also partake. Thus, in a classical music context, the generally accepted mid-frequency-occupied reverberation time for concert halls runs from 1.8 to 2.1 seconds. However, we can observe closely defined acoustic communities formed around particular concert halls—for example, Boston's Symphony Hall, with a reverberation time of 1.59 to 1.95 seconds; Vienna's Grosser Musikvereinssaal, with a time of 1.80 to 2.20 seconds; and New York City's Carnegie Hall, with a time of 1.57 to 2.12 seconds.[17]

Acoustic communities are bound together by shared evaluatory standards of acoustic conventions. However, to a member of a given acoustic community, acoustic conventions are simply recognized as natural or good sound. The acoustician Christopher Jaffe has observed that these musical-acoustic communities are "fiercely partisan regarding the quality of sound produced in their respective halls." From the example of the "explosive reaction of the New York musical community" to a "very minor physical change" to Carnegie Hall in 1966, Jaffe argues that a "narrower" "band for acceptable listening criteria" can develop within a community that is "acutely aware of the environmental characteristics of a specific space."[18] In other words, even though we have discussed acoustics from the onset of public concert culture, since Carnegie Hall opened only in 1891 and we can assume that the audience members who were upset over the changed acoustic had attended concerts at Carnegie Hall for an estimated maximum time of sixty years, the acoustic aspect of the figure of sound does not necessarily take long to establish.

In short, various competing demands have given rise to a particular configuration of audiences in relation to orchestras. For example, the demand that the performers be seen from the front gave rise to the convention that the orchestra is always in front of an audience that is facing them. As a result, the audience hears the music from a static position ahead of it, with sound moving only between left, right, and positions in between, according to which instrumental groups are playing. Paid numbered seats led to audiences always hearing music from the same relative space (the orchestra or balconies). The concept of the musical work gave rise to attitudes about careful and reverent intellectual listening and, in these attitudes' prosaic expression, to the convention that one does not move around the space or, say, dance during the concert. Together, these conventions limited the experience of music—and it was these limited concert-going experiences that gave rise to the spatial-relational and acoustic components of the figure of sound. Los Angeles's attempts to combine the criteria that emerged with the public concert hall with the concerns that an acoustic community (musicians, conductors, and audiences) have internalized clearly articulate the ossification of the acoustic figure of sound.[19]

Fine Tone Quality

This internalized figure of sound is conceptualized and verbalized in concert hall acoustics as "good acoustic" that allows for "clarity," "fullness of tone," and "fine tonal quality."[20] In the formulation of Leo Beranek, the "dean" of concert hall acoustics, fine tone quality is "the faithful transmission of the sounds from the instrument without any added (or subtracted) sounds, distortion, or shift of source."[21] Through interviews and research Beranek has defined a number of terms that form the shared language of musicians, concert patrons, and literature on concert and opera hall acoustics—although, as he acknowledges, these groups do not necessarily agree on them all.[22] Additionally, he has distilled his terms by systematically measuring the acoustics of fifty-four concert halls in sixteen nations (primarily created for Western classical music).[23] The result of "studied compromise," the list shows acoustical aspects that, over time, have come to be valued in relation to Western music performed in a closed space.[24] Furthermore, the spaces that Beranek considered, and that gave rise to his list, are concert halls and opera houses that seat more than 700 people—in other words, halls with a certain size requirement.[25]

1. Reverberation and fullness of tone
2. Direct sound, early sound, reverberant sound
3. Early decay time (EDT) (also early reverberation time)
4. Speed of successive tones
5. Definition (or clarity)
6. Resonance, including "Intimacy or Presence and Initial-Time-Delay Gap Liveness and Mid-Frequencies," "Spaciousness," and warmth
7. Listener envelopment
8. Strength of sound and loudness
9. Timbre and tone color
10. Acoustical glare
11. Brilliance
12. Balance
13. Blend
14. Ensemble
15. Immediacy of response (attack)
16. Texture
17. Echoes
18. Dynamic range and background noise level
19. Detriments to tone quality
20. Uniformity of sound in audience areas

Beranek has visually represented these aspects in a useful relational list (see figure 2.1).

Beranek's list shows acoustic phenomena communicated through a combination of scientific, musical, and architectural concepts and vocabulary. Jaffe provides an "Architectural Acoustic Translation System" moving distinctly among these three domains.[26] For example, what a musical conceptual system articulates as "Presence, brilliance, definition, transparency" is expressed in scientific terminology as "Arrival time of mid- and high-frequency reflections." In architectural vocabulary, these concepts would be expressed as "Location of boundary surfaces"; "Geometry of the hall"; "Audience to performer relationship"; "Relationship of the audience to reflective surfaces"; "Design of inner reflector systems."[27] Beranek's and Jaffe's shared goal is to develop a vocabulary that can capture and communicate concepts, so that the acoustics within the concert hall environment can be controlled to match the figure of sound. While Beranek is the first to acknowledge that his terms are not universally

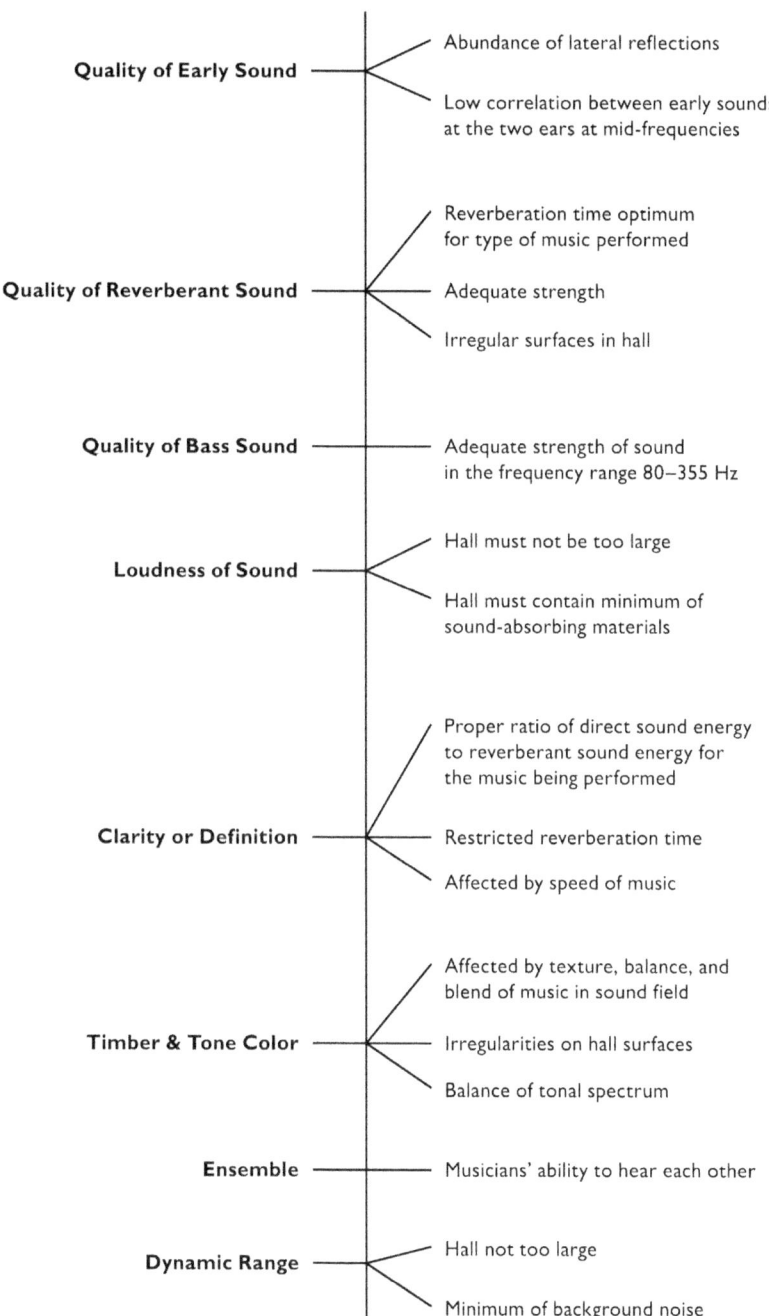

FIGURE 2.1 · This chart, created by Leo Beranek, shows the "interrelations between the audible factors of music and the acoustical factors of the halls in which the music is performed" (*Concert Halls and Opera Houses*, 34).

used and understood, centuries-old architectural and acoustic traditions have been internalized by audiences in cultures participating in these acoustic traditions.[28] Thus, while most audiences would use only a few of the nineteen terms and would be largely unfamiliar with the scientific terminology used to describe the acoustical phenomena, they nevertheless unconsciously evaluate most of the concepts as simply good sound versus bad sound.

The psychoacoustics of the ideal shoebox or natural sound constitute an interesting combination of a sense of directionality and clarity and a feeling of being surrounded by the sound. The result is a feeling of being immersed in a sound (rather than observing it from the outside or at a distance), yet the sound does not deteriorate into indistinctness. Fink summarizes the translation of this feeling into the scientific realm: "Both of these psycho-acoustic phenomena are linked to sound reflections which reach a listener from the side; *early lateral reflections* occur in the first 24–80 milliseconds, hitting the ears slightly out of phase to give a sense of what acousticians call *auditory source width*. Slightly later lateral reflections then contribute to a feeling of *listener envelopment*, and a more general bouncing around of sound waves from all sides finally creates the sonic 'bloom' of *reverberation*."[29]

While the acoustics of concert halls seemingly developed purely to serve musical needs, the concert experience that centers on public concert culture and the traditions built around this relatively new social form were also intimately tied to the conventions and traditions involved in visually based performances such as theater and dance. Thus, not only opera but even symphony halls have, to a large extent, adopted the visual frame provided around dance or theater.[30] As a result, the collective inner ear that we have developed to listen to music is tied to the visual/sonic image or situation of statically facing the orchestra while seeing and hearing the instruments in front of us, with the sound moving between our left and right: a static spatial-relational dynamic in relation to sound. Before recording technology this positioning mediated the majority of musical experience, and even with the advent of recording, stereo speakers re-created this relationship. Thus, this sonic/visual/visceral combination is inseparable from the sound, and sound is experienced within this spatial-relational acoustic package.[31]

Thus, in addition to air and normalized concert hall acoustics, we can add a third dimension to what constitutes the figure of sound. Because of the seating, which is pretty much constant, we develop a two-dimensional spatial relation to sound and its conceptualization. That is, there is a static relationship between the sound source and the listener. In other words, as audience members in a concert hall, we are generally seated in the same relationship to the music as our colisteners are. It is this relationship that is conceptualized and verbalized into discourse, and normalized to the extent that it goes unquestioned. Furthermore, the norm of sitting statically with the head facing forward also diminishes our acuity in hearing and sensing music as it takes place within three-dimensional space. The static, two-dimensional dynamic resulting from sitting in the same place for an entire concert, or even for multiple concert seasons — a peculiar dimension of the Western classical concert tradition — offers one possible acoustic dimension of that music. However, that two-dimensional acoustic dimension is naturalized. It is continuously affirmed and strengthened through sound production, reproduction, language, concepts, and historically and culturally informed perception of sound.[32]

In summary, we have been trained to feel that a reverberation time of about two seconds is normal. The naturalization of acoustics causes us to automatically (1) correct for gaps, (2) psychoacoustically auto-overcompensate, or (3) develop deaf ears and simply fail to hear any deviation. If we cannot carry out any of the above processes, we may believe the sound to be aesthetically erroneous, somehow wrong. Altogether, the process of naturalization and our listening within it creates a particular physical experience of good sound (the feeling that equates to the idea in the figure of sound). The ultimate outcome is that natural sound or sound that "stinks" becomes a screen to filter out, remove, or devalue all other sounds, leaving us with a limited concept of conventional aestheticized sound as stable, in front of us, moving between left and right, and having about a two-second reverberation.[33]

I will now examine two productions that break with the naturalized notion of acoustic-spatial relations. It is precisely because our sense of self, others, and our position in relation to the world is partially informed by acoustic-spatial relations that it is important to understand how common acoustic presentations affect both the music presented within them and how we hear it. Moreover, by examining any naturalized notions of acoustics, we also examine the ways in which the self, the other, and any possible relationships between them are naturalized.

Meredith Monk

Like Juliana Snapper's underwater opera, Meredith Monk's music escapes conventional analysis. As the *Washington Post*'s Anne Midgette puts it, "some dismiss [Monk's] music as simple or simplistic, but it is simple in the way that a tree is simple, with its hundreds of rings, its branches, its disparate leaves."[34] When, two decades ago, I first experienced a piece by Meredith Monk and

Ensemble, I was deeply affected, but I could not capture what it was about the concert that moved me. Considering it through concepts and vocabulary I knew from dance and music, I felt that the choreography was simple, and I could not articulate what exactly about the music captivated or had an impact on me. In fact, in many ways my reactions to Monk resembled my original reaction to Snapper's underwater opera project: for a long time I was not capable of understanding what was interesting about it, yet it continued to preoccupy me.

One reason for my baffled incomprehension might be that Monk's artistic work and expression are not limited by the traditional division between dance, theater, and music, a quality that has certainly met with challenges. While this topic is too large to discuss here, given the goals of this book, I will mention that Monk struggled greatly with national funding agencies, such as the National Endowment for the Arts, as it was unclear into which artistic category her work fell.[35] Thus, while Monk's music has been read thoughtfully from choreographic, theatrical, and operatic perspectives, in what follows I will posit that there is a different, multisensory way to understand her work. More specifically, I will suggest that if we limit our understanding of her work to one sense, that understanding will be greatly diminished. In other words, I will suggest that reading her music within the traditional acoustic aspect of the figure of sound limits our access to the artistic experience of the thick event. Specifically, considering Monk's work from a multisensorial perspective encourages inquiry into the way in which spatial-material relationships between sound-producing, listening bodies and the spatial-acoustic structures these relationships are organized within contribute to sound's affect.

Born into a family of musicians and educated at Sarah Lawrence College, Monk had become recognized for her work as a dancer and choreographer by the 1960s. However, even in her earliest solo efforts in downtown New York, she combined movement with images, music, and vocalizations.[36] In a quest to "make my voice move the way the body moves," Monk explored alternative ways of vocalizing.[37] She came to so-called extended vocal techniques through her own vocal experiments, her background in classical and folk music, and her experience as an "interdisciplinary performer."[38] With Laurie Anderson and Pauline Oliveros, Monk is often named as one of three female pioneers who have shaped the sound of American music. Over the years her work has been interpreted as choreography in which the dancers sing, opera in which the singers act and move around, and postdramatic theater with nonverbal vocalizing.

A major component of and resource for her subsequent portfolio, which so far spans four decades, was her establishment in 1968 of the nonprofit House Foundation, a moderately staffed organization that supports, produces, promotes, and maintains her work as a legacy. The House Foundation has produced a variety of work, from Monk's solo efforts to her operas and films, including the documentary *Inner Voice* (2008).[39] In 1978, Monk founded Meredith Monk & Vocal Ensemble. As the House Foundation preserves her work in institutional memory, the ensemble not only performs it but, since the music is not in notated form (with a few exceptions when it was transcribed much later), the members of the Vocal Ensemble hold, carry, and guard it within their bodies.[40] Although the core membership of the Vocal Ensemble has changed over the years, former members remain very much part of the life of the repertoire by teaching it and sometimes returning for select performances.

Not only do the Vocal Ensemble's members learn, memorize, and hold the repertoire within and with their bodies, but the compositions are also adjusted through, and in collaboration with, the bodies of these specific singers. Following is how some of Monk's longtime collaborators and performers describe the process.

Theo Bleckmann, a longtime key Vocal Ensemble singer, feels that working with Monk's music led him to singing from "muscle memory. A discipline I had never experienced before," because it was a process of "memorizing so deeply internally. Let go, just do it."[41] Allison Sniffin—who sings, plays the keyboard, co-arranges pieces, and has transcribed the few pieces that are now available in score form[42]—notes the creative relationship between Monk, the ostensible composer, and the performers: because "Meredith seemed to be okay with people finding themselves through her works," "I as a performer feel more and more free to put in sort of crazy ideas" rather than simply having to be "faithful to the first improvisation she did." As a result, "we do a lot of that work in collaboration."[43]

The soprano Katie Geissinger reflects on the roles and importance of individual singers in the formation and cohesion of the ensemble: "What I realized was that she was picking me for me, and not for skills that were put on, inhabited by me. And that has really helped me."[44] The process described by these three key performers is akin to a site-specific piece: it would not be the same were it to be performed with and through the body and voice of another singer. Correct execution of sound or movement does not constitute the piece; instead, the piece lives and exists through a particular person's materiality and life story. And it is through the challenges of the piece that vocal aspects the performer has not previously had a chance to explore are engaged. In other words, both performer and composition gain invaluable and unique material from one another and contribute uniquely to each other.

Songs of Ascension

Having followed Monk's career since attending the concert mentioned above, in 2008 I finally understood what intrigued me about her works: the issue of compositional and performative spatial specificity, which she contemplates throughout her oeuvre. In addition to composing within the parameters of sound and time, Monk also composes for the spatially specific relationship between the sounds, as well as in terms of the sounds' overall relationship to the space. Reflecting on Paul Celan's poetry about the songs of ascent, a series of psalms sung by ancient pilgrims as they climbed a holy mountain via a series of steps, Monk created a movement, light, voice, string quartet, and percussion piece that embodied the energies and dynamics of the circumventional and vertical movement axes. Hence, *Songs of Ascension* (2008) was a response to questions Monk asked herself: "What did the songs of ascent sound like? What did the voices sound like?"[45] She found this practice, and the general idea of reaching upward—emphasized in several spiritual practices—fascinating.

As Monk was working on *Songs of Ascension*, the visual and installation artist Ann Hamilton invited her to sing at the opening for one of her sculptures. The sculpture in question was an eight-story work in the form of a tower with a stair, situated in the private sculpture garden of the Oliver ranch in Geyserville, California (see figure 2.2).[46] Monk describes the experience: "We're on parallel staircases, but we can never reach each other." The stair, modeled on a double helix, creates the illusion that you can touch the people on the other side. Similarly, there was a relationship between closeness and distance, in that "the audience couldn't see the entire thing [the performers], but you could hear the whole thing [the music]." Furthermore, Monk notes that "the acoustical situation in that tower was so unique and, you know, to be able to even hear each other was so interesting." In terms of performer placement, Monk was "trying to work with that very extreme, like sometimes something would come from way down at the bottom, and the rest of the performers would be way up at the top and one performer would be way down at the bottom."[47]

Hamilton's tower became not only the first of many memorable performance spaces to host Meredith Monk & Vocal Ensemble, but, with its ascending architecture and weaving stairways, it became *Songs of Ascension*'s physical and compositional touchstone.

FIGURE 2.2 · Meredith Monk's *Songs of Ascension* performed inside the sculptural tower created by Ann Hamilton, Oliver Ranch, Geyserville, California (photo by Maria Mikheyenko).

After the concert at the tower, the piece was brought into a number of new performance contexts. In each new setting, *Songs of Ascension* was mounted in response to the space. As a touring art show is uniquely laid out and configured for a new situation, Monk adjusts the piece spatially in response to a given hall. By *mounted*, then, I mean not that the piece is just brought inside the space as a static object, the same each time independent of the space, but instead that it is configured for, and according to, the specific space. While the overall choreography among the singers is retained, necessary adjustments are made in response to new spatial conditions. In this way, the piece is as much about reconstructing and creating a sculptural space as it is about the particular notes that are sung. Hence, when entering a new performance setting, Monk asks what the space tells her. She then remounts the piece as needed, with the aim of offering an immersive, circular experience for the audience in venues quite different from the first performance in the double-helix tower.

Monk thinks about the process of creating *Songs of Ascension* as a return to her ideas about site specificity, which she nurtured during the 1960s. As she mounts the piece in different settings, I hear her choreographic and spatial adjustments as reorchestrations that respond to the space through the music. Through this process she not only throws the space into relief, but she also allows the space to call forth a new aspect of the composition. *Songs of Ascension* has been brought to such diverse places as the Great Hall at Dartington College of Arts, Devon, England (May 13, 2008); the Memorial Auditorium at the Stanford University, Palo Alto, California (October 18, 2008; see figure 2.3); the McGuire Theater in the Walker Art Center, Minneapolis, Minnesota (June 12–14, 2008); the REDCAT Theater in Los Angeles, California (October 29–November 2, 2008); the Harvey Theater at the Brooklyn Academy of Music, in New York (October 21–25, 2009); the Walt Disney Concert Hall in Los Angeles (April 11, 2010; see figure 2.5); and the Solomon R. Guggenheim Museum in New York City (March 5, 2009; see figure 2.4). The piece has been transposed into spaces that included black box theaters; traditional concert halls; elevated stages; performance spaces on the floor; and three-dimensional spaces, where the performers move not only in the two traditional directionalities—stage front and back and stage left and right—but also move vertically. In the majority of these spaces the audience was seated in one place and position in relation to the piece throughout its duration.

The sonic particularity of each performance space derives in part from its unique architectural, material, and spatial-relational properties. However, much technical work on concert hall acoustics aims to unify or homogenize the sonic experience from each seat in the house, undermining the facts that seats in different sections exhibit different physical and spatial relations to each sound source, and each sound source relates in its own way to the hall in general. In contrast, by making considerable choreographic (or acoustic-relational, as I see it) and sonic adjustments—which I have termed reorchestration or transposition—to *Songs of Ascension* each time it is performed in a new space, Monk recognizes and responds to the spatial and relational specificity of each situation, and indeed of each encounter that takes place during singing, listening, sound, and music.

Hence, in the same way that Monk composes not only for, say, a soprano voice but for a specific person in her ensemble, *Songs of Ascension* was created for a specific spatial-acoustic situation. As a result, when it is performed in a different spatial-relational situation, we could make the analogy that Monk has a different set of instruments available and so needs to reorchestrate the work. In other words, the piece would lose its identity if it were not adjusted to the spatial-relational dynamic between the various performers and their overall relationship to the hall. This shows us that, in each remounting of the piece, Monk must respond to the dynamism of the hall. Acoustics is such a crucial part of sound's identity that Monk has to respond to the specific acoustic conditions of each new performance space.

Similarly, Emily Thompson's extensive aural history of early twentieth-century science and engineering demonstrates that how we think about natural sound or just sound has, in fact, been carefully manufactured through isolation and abstraction from its original acoustic context. Indeed, Thompson points to an underlying commitment to an idea of sound that "had little to say about the places in which it was produced or consumed."[48] Indeed, by 1932, innovations in electrical engineering and acoustic design were used to strip sound of the actual sound of space. Sound of space was now an element that could be "added electronically to any sound signal in any proportions; it no longer had any relationship to the physical space of the architectural constructions."[49] In the construction of concert halls, extrasonic concerns influenced the architectural layout—and the acoustic design needed to compensate to hold the notion of (metaphorically speaking) pure sound in place. Essentially, concert hall design is a numbers game in which money and prestige fight against the laws of acoustics, architectural aesthetics, and seating politics. How can we maximize the ticket revenue for each performance? Who gets to sit where, and for what price? Presenting music in spaces of commerce—that is, concert halls with purchased seating—has pinned us to the mast (echoing the previous chapter, I again invoke Max Horkheimer and Theodor Adorno's critical imagery).[50] Concert hall acoustics attempts to create the ideal sound from

FIGURE 2.3–2.4 · Above, top: Rehearsal of Meredith Monk's *Songs of Ascension* at Stanford University (photo by Maria Mikheyenko). Above: Performance of Meredith Monk's *Songs of Ascension* at the Solomon R. Guggenheim Museum (photo courtesy of Stephanie Berger/The New York Times/Redux).

FIGURE 2.5 · Performance of excerpts from Meredith Monk's *Songs of Ascension* at the Walt Disney Concert Hall (photo by Axel Koester).

every seat, working to erase sonic spatial specificity in favor of predictability, consistency, and profit. In contrast, Monk's twenty-first-century operatic work on sound in, through, and as space shows us the irrationality of the proposition that we can sense and relate to sound divorced from its spatial characteristics.

Based on the different events I attended, I hear *Songs of Ascension*'s performances as falling into two different experiential types. In the first type, as discussed in detail above, Monk remounted the piece in new halls, and within those configurations, each new iteration took on a slightly expanded identity compared to the first performance situation. I believe that Monk's spatial reconfiguration shows us the spatial-acoustic specificity of sound. Through her remounting of *Songs of Ascension*, Monk attempts to denaturalize the idiosyncratic spatial specificity that the economy surrounding Western concert halls strives to mask. In the second type of experience, Monk was not able to retain *Songs of Ascension*'s pitch-time-space dynamic. While she made acoustic-relational adjustments for this performance, *Songs of Ascension* seemed very different in its Walt Disney Concert Hall incarnation — very small and disconnected from the space and situation. Compared to my past experiences of the piece, this one sounded off to me. However, because — based on past experiences — I doubted whatever I thought about this performance of the piece, I began to doubt the concert hall. Yet I had also had experiences in which this hall sounded good. This led me to consider the intersection between the piece and the concert hall. Indeed, Monk not only challenges the figure of sound, but she also makes us aware of it by composing outside the established spatial-relational acoustic norms.

Yet I wonder whether it is also partly Monk's continued presentation of her work in spaces where three-dimensional dimensions are explicitly considered that has created an acoustic community able to hear her work with that sensibility, making the gap clear rather than automatically failing her work within these idealized spaces. On some levels, Monk's devoted audience has developed a familiarity with and an appreciation of an alternative sonic system — which is, of course, simply another figure of sound. And with that alternative sensory-sonorous familiarity comes the ability to comprehend and accept both two- and three-dimensional acoustics and their dynamic relationship to each other. That is, rather than praising or dismissing Monk's music as superb or inferior, her audience can sense that it is placed in its native acoustic environment (rather than sensing that it requires three-dimensional acoustics but has been placed in a two-dimensional acoustic space). This allows for a kind of evaluation that takes place not on a positive-negative scale, but rather through the understanding that any positive or negative perceptions have to do with the alignment or misalignment between the acoustic dimension for which the music was written and that within which it is played out. In this way, Monk's music raises questions about the relationship between the sound and the system of the figure of sound. The question suggested by such a position is: Is this right?

Rather than considering Monk's inability to retain *Songs of Ascension*'s pitch-time-space dynamic in the Walt Disney Concert Hall performance as a failure, I believe it shows us a crucial facet of the acoustic aspect of the figure of sound, and the way that the space within which we hear music inextricably participates in forming the figure of sound as well as our experience of that figure — a facet rarely allowed to come forward. As the intense discussions about the building of the Disney Hall illustrated, classical musicians rarely play outside the context of normal, beautiful acoustics. And the particular music played in rarefied concert halls rarely requires different acoustics. The relationship between singers and compositions may be viewed as parallel to the relationship between compositions and halls. In the same way that the singers in the Vocal Ensemble found themselves stretched and challenged by the compositions, the compositions are highlighted and the performers' specificity is brought into being, new aspects of the compositions come forth in new spatial-acoustic settings, and aspects of the hall are redefined as it sounds material that is not written directly for the acoustic ideal it embodies.

So while we could read Monk's performance as a composition written for specific acoustic conditions inserted into inappropriate acoustic conditions, I believe there is much more to glean from this story. Through realizing that the piece sounds better or worse depending on the hall and its flexibility in terms of performer-and-audience spatial configurations, and that the Disney Hall sounds sometimes better and sometimes worse depending on the music it reveberates, the careful listener realizes the extent to which the encultured and sensorially spatial-acoustic aspects of the figure of sound are deeply engrained. They are so engrained that, rather than considering the dynamic relationship between music and its acoustic conditions, listeners often fail the music or the concert hall. This process is poignantly captured in Frank Gehry's summary of the acoustician Minoru Nagata's assessment: "Berlin stinks!"[51]

It is, in part, by investigating music from a multisensory perspective that we may account for dimensions of music that are known but frequently overlooked by analysis. As we can recall from chapter 1, viewing Snapper's work through this lens illuminates the fact that sound does not exist in a vacuum but rather is always already in transmission; its character therefore arises from the material particularities of each transmission. Applying the multisensorial perspective to Monk's work highlights the notion that a sound's affect depends on the spatial relationship between the sound, listening bodies, other sounds, and the larger situational canvas on which these relationships take place.

Invisible Cities

Another production of an operatic piece that opened my senses to additional aspects of the figure of sound took place five years after I first attended a performance of *Songs of Ascension*.[52] Opening on October 19, 2013, this was a production presented as "an invisible opera for wireless headphones." I concentrate here on the production of the piece rather than on the content and music of the opera, *Invisible Cities* (based on its namesake, Italo Calvino's 1972 novel), composed by Christopher Cerrone. The opera company The Industry and the dance company the LA Dance Project, both located in Los Angeles, mounted the production.

Founded in 2011, and with *Invisible Cities* marking its second major production, The Industry has already made an impact on Los Angeles's new music and opera scene. Its dynamic director, Yuval Sharon, was formerly director of VOX: Showcasing American Composers, the New York Opera's new opera program, and he has directed or codirected traditional and new opera at venues including the San Francisco and Los Angeles Operas.[53] Thus far, The Industry's work has been marked by an intense engagement with issues related to the continuing transformation of the Los Angeles region and Southern California in general. Part incubator for emerging talents, part collaborator with established artists, its productions are marked by experimentation and engagement with emerging technologies, with the goal of breaking the bounds of the proscenium.

During an interview with KCET, a Los Angeles television station that did an hour-long documentary on the production of *Invisible Cities*, Sharon commented that its concept resulted from a challenge by the sound designer E. Martin Gimenez, who dared Sharon to consider "an opera for headphones."[54] Using microphones, sound design, and headphones in the context of live opera performance is not a casual technical decision. It is a philosophical and, some would say, moral decision, as historically and fundamentally, opera owes its very sonority and existence to a feat of acoustic virtuosity. Singers' voices are pitted against the sounds of the orchestra and challenged to rise above them to reach the audience. All singers can rely on is their decades of training and the acoustic condition of the house. In this context, the unamplified power of singers' voices is part of the fetish that defines the art form.[55]

Sharon took on Gimenez's challenge. To take the concept to its fullest, Sharon realized, he wanted to set the opera in a public space, a situation that would "blur" the line "between everyday life and art."[56] On first blush, that setting and mediation of the piece share characteristics with silent disco and the Metropolitan Opera's high-definition video transmission, *Live in HD*. As in silent disco events, the audience would inhabit two spaces: the physical space and the space provided by the sound emitted from the headphones.[57] However, unlike some silent disco gatherings in which people bring their own music and the primary sharing takes place through listening and dancing together (to different music), the audiences of *Invisible Cities* shared not only the music, as in the Metropolitan Opera's high-definition video simulcast, but also a live performed music, albeit primarily experienced via headphones.[58]

For Gimenez, whose original challenge to Sharon initiated the opera-for-headphones endeavor, "sound design is as much a character as the music."[59] Studying the libretto, Gimenez designs the sound to communicate the drama, just as the director and lighting and costume designers do. To Gimenez, the "sound design is going to be as much of a character in the piece as the text, as the singers, as the dancers." For the sound design for *Invisible Cities*, Gimenez was thinking in cinematic terms. He explains his thinking process about the sound design for the opening: "Kublai Khan [is] alone in this Palace." The libretto begins: "There is a time of emptiness that comes over everything."[60] Taking on the challenge of conveying the character of "emptiness" in this scene, Gimenez asked himself: "How can I create [the camera's] close-up to a very wide angle" in sonorous terms?[61]

The solution was to render the voice "bone-dry for that first line. . . . And then sonically, over the first line, over a minute," Gimenez explains, "we kind of sonically pan out, and this cathedral reverb slowly fades in and you kind of realize 'Oh, wait. He's all alone in this vast space.'"[62] Acknowledging that because traditional "opera is based on hearing things unamplified in a beautiful room," while *Invisible Cities* goes "to the extreme opposite," Gimenez reflects that, under his design, "each movement, each line kind of has a sonic character [related] to that." He asks rhetorically, "How do we achieve that sonic character?" Answering his own question, he says: "Using ambient mics [sic]. Using a lot of fake reverb within our console. That will help us to determine, dramaturgically, the goal within each scene." Using microphones and digitally determining the voices' reverb and placement in space shifts the aesthetic premise and value, traditionally bound up with singers' ability to both ride the room's acoustic and train for years to gain the vocal power necessary to match an orchestra and fill a space the size of an opera house. What you gain in a digitally controlled situation, in Gimenez's opinion, is dynamic range in the lower end (for example, the ability to communicate in a whisper) and the ability to change the reverb and sound placement, and hence imbue these parameters with meaning. As the production's tagline quoting Calvino says, "It is not the voice that commands the story: it is the ear."[63] As I discuss below, with the opera sound designed and delivered via headphones, the ear cannot detect spatial specificity and, hence, cannot lead the audience to the opera's actions. Perhaps the ears referred to are Gimenez's and Sharon's, and their offering of what they are hearing through production and design?

In response to these conceptual frames, the production is set in Los Angeles's historic and iconic Union Station, which most people have seen as a backdrop for such movies as *Blade Runner*, *Pearl Harbor*, *The Hustler*, and *The Dark Knight Rises*.[64] It was therefore in these worlds—fantasies that, in Los Angeles and around the world, are often given value and allowed substantial room in people's emotional lives—that the opera was set. Additionally, because Union Station is nestled between downtown, Little Tokyo, and the Flower and Fashion Districts and barely out of the shadow of the neighboring Los Angeles Central Jail, multiple layers of the city are already invoked and activated for and by audience members before and after their arrival.[65]

My description of the 2013 production of *Invisible Cities*, which follows, is based on my attendance at two performances (the first time with headphones and the second without them), and on my engagements with performance recordings and video documentation footage, interviews with the producers from a television documentary about the opera, and press and media coverage. In general terms, the piece consisted of solos, duets, trios, and ensembles. From my vantage points as an audience member, the activity moved from the general position of the South Patio to the North Patio and then into different locations in the main waiting hall and main entrance area (see figure 2.6). Some performers also started singing in one of the Patios and then moved into the building while singing. Voices that sounded together on the headphones were not always singers singing physically together, and, for one person attending only one performance, it was challenging to get an overview of all the voices' physical placement within the space.[66]

Experiencing Sharon's and Gimenez's Production of *Invisible Cities*

On the evening of the performance, audience members arrive at the station like other commuters: by train, metro, car, bus, bike, or on foot. But after the audience members make their way through the walkways that lead from park-

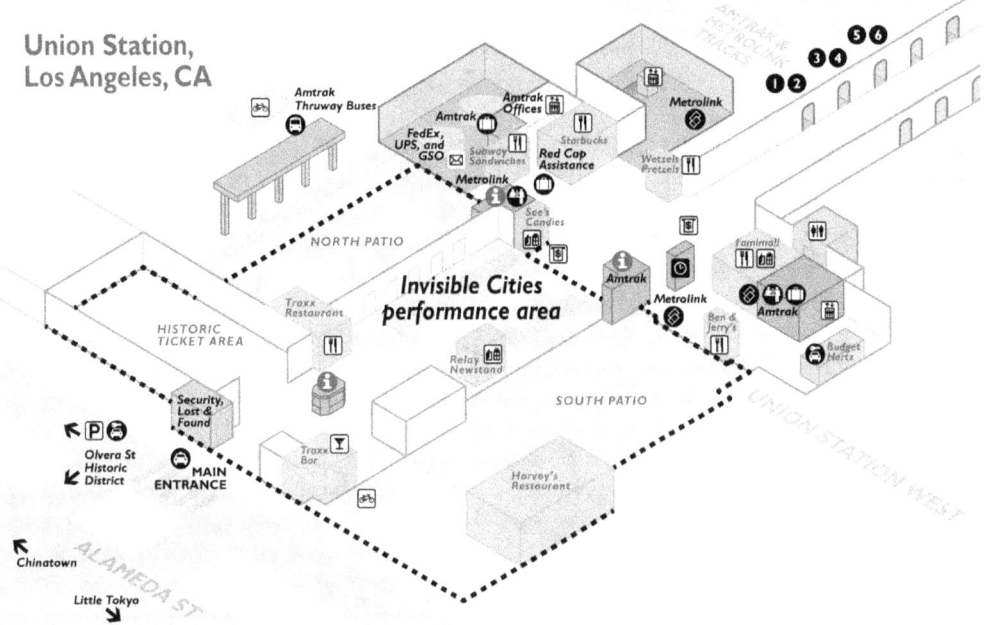

FIGURE 2.6 · Map of Union Station, Los Angeles, California (drawing by April Lee).

ing garages and train tracks, or from one of the two main entrances, and reach the main waiting hall, nothing looks or feels different from any other evening at the station. The only minor change is that the historic ticket lobby, which is normally closed to the public, is open for those with tickets to the performance. Ticket holders pick up a set of headphones at the original ticket counters. While activity in this area of the complex is not part of the station's daily life, commuters are already familiar with the area's popularity as a location for special events and filming. Indeed, the performance that is about to start is not noticeable.

As curtain time nears, ushers in everyday clothes, wearing small pins marking their affiliation with the event, direct audience members to the Harvey House Restaurant, the iconic venue that has not been in general operation since 1967 (see figure 2.7). On the restaurant's main floor, cleared of tables, the chamber orchestra is installed. People crowd around the edges and into the restaurant booths along the walls. Sharon welcomes the audience and offers basic instructions, mostly precautions regarding cohabitation with the station's regular occupants. The event starts when the orchestra begins the overture.[67]

The Acoustic Mediation of Voice, Self, and Others · 83

FIGURE 2.7 · The overture of Christopher Cerrone's *Invisible Cities* played in the Harvey House Restaurant, Union Station (photo courtesy of The Industry).

Audience members stay for a while, watching and listening to the orchestra, before leaving the space of their own accord. After exiting the restaurant, where the orchestra was both heard and seen and the sound was fairly similar with or without headphones, each audience member's experience of the opera takes a unique path. That is, for each audience member, the opera unfolds according to his or her specific sonic, visual, and spatial experience. Accordingly, I will now adopt a first-person narrative, reflecting one iteration of the opera as it took place from a single perspective.

Leaving the Harvey House Restaurant, I am led directly into the enclosed South Patio. I see some people with headphones forming clusters and moving around together, while others move alone. At first there is only orchestral music coming from the headphones, with the vague hum and bustle of the station pressing in, and sirens filtering in from a distance. As soon as I hear a voice in the headphones, my inclination is to go and find its source. It is when I hear sounds that I do not see, when I cannot tell immediately even from which general direction they come, that I begin to sense the gap between the acoustic and sonic world surrounding me and the omnisonorous sound world offered through the headphones.[68] Since the acoustic cues conveyed by the mixed music do not reflect the music's placement in the physical space and acoustic character of the station, I find myself relying on visual cues such as gatherings of small crowds to seek out singers' locations. I assume audiences have gathered around the activity I hear.

The first performers I see are dancers in the South Patio, dancing in spotlights lighting up the garden (see figure 2.8). As everyone who was in the Harvey House Restaurant during the overture has to move through this area, it is so crowded that I move into the main waiting hall, heading from there toward the areas that lead to the gates and tracks. Before I even make it to those areas, two friendly people whose pins identify them as ushers emerge to let me know that the performance space ends at the edge of the waiting area. From that point until the finale, I move between the North Patio, the main waiting hall, and the area near the main entrance (on the Alameda Street side). In this way, audience members are subtly directed through and dispersed throughout the space, drawn to particular areas by an activity or away from others by ushers.

The overall concept of an invisible opera emerges not only from the performers' engagement in the everyday activities of the hall, such as sitting down in the waiting area and reading a paper or cleaning the floors, but also because the performers are, both for rehearsals and performances, situated among the station's everyday patrons—both travelers and those who use the building for shelter (see figure 2.9). It is only when they break into operatic-style song that the singers identify themselves (people frequenting the station sometimes sing as well!). In some ways, the line between the dancers and the patrons is arguably less distinct than that between the singers and the patrons. Many of the patrons are highly creative in dress and movement, and before I begin to notice repeated dance vignettes and recognize specific dancers, my occasional difficulty in identifying them illuminates the blurring of the line between everyday life and performance and the emerging and receding of this particular performance. And, perhaps as importantly, seeking out performance or heightened moments throws everyday life into relief as performance or as art.

Wearing the headphones distinguishes audience members from the station's everyday patrons. In contrast, the singers are indistinguishable from the patrons in terms of appearance. I find myself playing a silent guessing game, wondering whether a particular extravagant-looking person is part of the cast. As the different singers start to sing throughout the performance, I realize that some of them wear everyday clothes while others are more theatrically marked. However, some of the people who act, move, and appear theatrical in the station are not part of the cast. Again, while there is no obvious distinc-

FIGURES 2.8–2.9 • Above, top: Dancers in Union Station's South Patio during a performance of Christopher Cerrone's *Invisible Cities* (photo courtesy of The Industry). Above: On-site rehearsal of Christopher Cerrone's *Invisible Cities* (photo by Dana Ross).

tion in appearance—some nonperformers are more theatrical in outfit, body language, or vocality—as an audience member I begin to distinguish between those voices that are mediated through the headphones and those that are not.

With the ominous headphone sound following me, I wander around, having to constantly decide whether or not to move toward the areas where I see that other people wearing headphones have gathered (figure 2.10). While sound is normally the cue to move toward a performance, with the curated sound in the headphones—which is the same for every audience member regardless of his or her physical relation to the sound source—sound is no longer a cue. Instead, a sonorous space is created in terms of the opera's voices and orchestral sounds. Their sound design puts them in relation to one another, but they remain outside the usual physical and spatial relation to the audience members.

For patrons of Union Station, only the singers' acoustic voices are audible. The singers wear tiny lavalier microphones and in-ear earphones. The orchestra and vocal sounds are mixed live and returned to both performers and audience. While the orchestral musicians can see the conductor at all times, for the singers the earpiece carries their only cue. As noted earlier, audience members wearing headsets are exposed predominantly to the designed mix of operatic voices and orchestra. While volume can be adjusted individually, the headsets also function by default as light sound mufflers, generally limiting the sounds of the station's activity to those that pierce through during loudspeaker announcements.[69] I frequently see audience members taking off their headsets, as I do myself at times.[70] Sometimes people remove their headphones to share the sound with a friend or passerby. At other times, it looks like audience members simply remove their headphones for a while to take in the acoustic sound.[71]

I also observe that this operatic performance in a public space carries with it an invisible and amorphous separation between audience and performers, yet barriers are broken down between the audience and station patrons. The audience keeps a respectful physical distance from the performers yet feels complete license to observe them. At the same time, their common reference point—the opera—contributes to partially breaking down the invisible barrier that often exists between strangers in a public space. I observe more conversations and interactions between strangers than on an average day at Union Station or in another public space. Throughout the piece, the performers inhabit all the different public areas of the main hall and the two courtyards. Only the very last scene separates both performers and audience from the patrons' bustle, moving performers and audience into the original ticket lobby that is normally not used. (I will discuss this scene in more detail below, but it is

FIGURE 2.10 · Performance of Christopher Cerrone's *Invisible Cities*, main hall of Union Station (photo courtesy of The Industry).

useful to first discuss the composer's and director's decisions to use amplified sound, headphones, and sound design.)

While the production of the piece encourages the audience to engage with the acoustic world offered through sound design, *Invisible Cities* cannot fully mask, or offer a clear sense of, the acoustics of the hall.[72] By creating tension between two simultaneous acoustic worlds, the production exposes naturalized acoustics as part of the figure of sound. The opera's climactic ending—in which, for the first time, all of the singers and dancers are gathered in the same space—illuminates the two simultaneous acoustic worlds and hence the gap between them. Throughout the opera, the singers' voices have been mixed by Gimenez and presented to the headphone-wearing audience as a single sonic mix, while the audience may see none, one, or only a few of the singers. Because of the gap between what is experienced live and what is heard through the headphones, the music presented via the headphones can feel in many instances like a moment of listening to recorded music in a different sonorous situation. However, because during the ending the audience can see all of the singers they hear, this illusion is not sustained. This moment marks the first time the singers' acoustic realm merges with the material realm—that is, the audience members see what they hear. But while some sense of stability may be provided by the parallel between the number of performers seen and heard, the gap between the two acoustic realms—the sound design and the live acoustic of the ticket hall—may arguably be more unsettling.

This also means that, during this last scene, the audience can alternate between listening with headphones and listening without them. The difference in acoustics creates an opportunity to compare the two experiences and either reject the headphone or live experience as inferior (failing to match the figure of sound) or accept the headphone or live experience as valid (whether as the figure of sound or as something else that is just as useful as—or even more useful than—the standard idea of sound). In this way, the simultaneous physical inhabitation of one acoustic space and aural inhabitation of another shows the power of sound's acoustic space and context, while laying bare the acoustics of the voices as they sound in the train hall. We may recall that the shared acoustic space offered by the headphones "creates a bond between you and the other headphone-wearing audience members."[73] In the last scene, the two simultaneous acoustic worlds rub up against each other, showing us how much they form the sound as we hear it. In other words, the acoustic, spatial-relational world is inextricably bound up with the figure of sound.

As is the case with sounds' transmission through air and bodies, we are not often party to the way in which a spatial-relational acoustic is naturalized. Because we cannot help but hear sound within an acoustic wrapping, it is difficult to gain the insight offered by the mismatching of actual acoustic space and the acoustic space offered up through headphones, as created through the production of *Invisible Cities*.[74] And, as in the previous chapter, this situation initially made me cringe because I first heard it through the first option described above. Therefore, while my initial reaction was that it was a pity the production had used headphones, the hypercontrolled acoustic in the otherwise non-operatically attuned structure and the readily available option to experience both (by wearing or removing the headphones) turned out to be a wonderful place from which not only to discuss the hypermediated acoustic nature of sound and music, but also to understand the situational-relational and acoustic aspect of the figure of sound.

Accoustics and Listeners' Choice

But if acoustics is a naturalized part of the figure of sound, would that not imply that we cannot hear outside the naturalized — and, by definition, that we cannot attain the distance necessary to sense that it is indeed naturalized? That is, does a naturalized acoustic imply that we are enclosed in it, and that we do not have options when listening? *Songs of Ascension* and *Invisible Cities* negotiate this issue by being simultaneously inside and outside. In other words, they are not so foreign that they are dismissed as complete anomalies. To me, the interesting thing about the productions discussed in this chapter is that they, in very different ways, work within the model of the figure of sound while also resisting it. By composing without the constraints of the two-dimensional model of the musical work and music making, Monk resists the two-dimensional naturalized notion. However, by also allowing the work to be presented in halls that specifically accommodate and promote the two-dimensional figure of sound, Monk exposes the acoustic condition of these halls, a condition that is simply understood as normal. Under these conditions, "musicians can play without undue effort, and without the annoying adjustments of balance that are all too essential in many of our concert halls today."[75] Thus, audiences must decide whether or not to simply judge that the piece is poor. *Songs of Ascension*'s presence in the hall does two things: it points to the figure of sound by showing that we hear and make sense of music through it; and it allows us to become conscious of the measurements with which we evaluate the hall as a vehicle for delivering that figure of sound.

By providing a sonorous envelope and an acoustic landscape that are familiar to us in opera (and also film, since the acoustic conditions in *Invisible Cities* changed throughout as they would do in a film, but not in an opera house) within an entirely different acoustic condition—a train station where we are sometimes in earshot of the singer and sometimes not, and nearly always outside of the orchestra's live sound range—*Invisible Cities* offered an entirely different window into the ways in which the acoustic condition is written into the figure of sound. To put it another way, *Songs of Ascension* brought a three-dimensional piece into a two-dimensional concert hall acoustic, while *Invisible Cities* brought concert hall acoustics into a live, idiosyncratic acoustic space that was vastly different in its physical and acoustic relationship to the operatic performance. In this way, *Invisible Cities* is an extreme example of the acoustic dimension of the figure of sound, as it is played out digitally. While this acoustic is extreme—or, arguably, postacoustic in being aided by microphones, mixers, digital reverb, and headphones—it is only a heightened version of what is otherwise designed for and manifested within the acoustics of concert and opera halls.

To me, then, what is important about these works is that they present in-between points that do not reject the normative acoustic world, but rather are outside it enough to allow us to sense what is naturalized. By putting the naturalized and denaturalized dimensions together, they embrace, deal with, or play within the gap, allowing us to sense it (which something that exists comfortably within a naturalized acoustic does not allow us to do). In the same way, despite the strength of the spatial-relational and acoustic aspect of the figure of sound, *Songs of Ascension* and *Invisible Cities* show us that there are ways of playing within the acoustic norms just so—just enough to venture only slightly outside. The effect of an experience that exists largely within the norm is not so unexpected that it will be discarded or disregarded. It is precisely through refraining from leaving the naturalized dimension dramatically, and thus remaining within what is considered as opera or music, that these productions offer the potential to experience the naturalization.

Conclusion

This chapter has offered insights into two aspects of sound and perception of sound—the acoustic element and the spatial-relational element of sound—both of which are inextricably bound up in the figure of sound. From this we may conclude that, at any given time and context, we cannot but hear and make meaning with sound and music as they unfold within or challenge acoustic and spatial-relational norms. Through *Songs of Ascension*, Monk shows us that sound is naturalized in spatial-relational acoustic terms. Through the

Union Station production of Cerrone's *Invisible Cities*, Sharon and Gimenez show us that at least two overlapping acoustic systems can more than coexist. That is, in the same way that air has become the naturalized medium through which sound is transmitted, the spatial-relational and acoustic norms of concert halls contribute to the constitution of the figure of sound. As I discussed in chapter 1, we hold a concept of the pitch A (440 hertz) sounded for one quarter-note in the tempo of sixty, or andante, specifically as it is sounded when transmitted through air. Additionally, included in the concept of that pitch in a normalized mode is directionality (it is sounded in front of us) and acoustics (with approximately two seconds of reverberation, and meeting all of the other acoustic criteria discussed above). Moreover, the sound is probably relatively stable in its relationship to the person who imagines it—that is, neither the person nor the sound moves relative to each other. In summary, expectations and norms are set not only about distinct sounds, the work, and its performance, but also about the acoustic conditions in which the work is presented and experienced.

Such standardization perpetuates itself, and variation is understood as anomaly: composers know music within this acoustic and write for these conditions. Similarly, audiences have internalized sound and music within such acoustic parameters and, as the above vignettes exemplify, feel a sense of disconnection and wrongness when these conditions are not met. While the acoustic conditions discussed in this chapter arose from a particular and limited musical culture and experience, it is this musical culture (and the specific experience it offers) that has come to dominate language and concepts about, and epistemologies of music within, Western culture. That is, while Western culture's music—and thus the acoustic experiences it offers—is far from limited to the specific acoustic condition discussed here, it is this precise acoustic condition that we have used to set the acoustic norm. As a result, other acoustic conditions (even those that resist the norm) have primarily been understood in (negative) relation to it.

I want to stress that, for denaturalization to take place, a listener must experience and actively participate in it. *Songs of Ascension* reveals the relational acoustic dimension of the figure of sound, opening up the possibility for understanding that how we evaluate sound is based on listening within systems. I aim to change the question "Is this sound good or bad (or right or wrong)?" to a different question: "Which system is this sound created within, and which system is it experienced within that makes it appear good or bad (or right or wrong)?" The realization of *Invisible Cities* in Union Station, however, nudges the question one step further. Instead of asking "Is it right?" or "Does it seem right?," the production returns to the primary question: "What is it?" If asked honestly, this shift not only opens up the possibility of understanding an additional sonic or acoustic system. Because the new question is not directed toward the identity of the sound, it is not tied to a known listening system or dimension—which allows for a different mode of listening to opera and voice.

As I have discussed, the spatial-relational aspects of the figure of sound are naturalized through reifying practices in architecture, acoustics, composition, concert programming, and musician and audience pedagogies. Because of this deep reinforcement, moving the intellect is not enough. For sound and music to be denaturalized physically and experientially, we need to diversify and enrich our sensory abilities, giving ourselves a wider range and depth of sensitivity to that which we refer to as sound and music. It is precisely because the work takes place in the sensory dimension before it is used as a basis for meaning making that deflation of the figure cannot take place through intellectual power only.

Recall my initial reaction of discomfort and embarrassment about the music because it sounded so awkward. By examining this phenomenon further, I came to understand that the awkwardness was due to my encultured listening to the two pieces.[76] Rather than dismissing the pieces and the particular performances, by considering them more closely I was able to understand that my first perception that the music was lacking (was wrong) was not about the music per se. Instead, I was sensing the gap between the acoustic aspect of the figure of sound, which I had internalized, and what was sounded.

It is precisely because they differ only slightly from pieces that would fit squarely within the naturalized mode that *Songs of Ascension* and *Invisible Cities* can expose the norms as norms only. Because the pieces remain within the frame of the normative acoustic configuration of sound, audiences can still accept them as normative music (whether judging them good or bad). As a listener who felt discomfort and embarrassment, then, I was presented with a decision point: to engage through the unfamiliar, nonnormative spatial-relational and acoustic aspects of sound that were actually present, or to dismiss the situation because it did not fit into known models. It is because *Songs of Ascension* and *Invisible Cities* violate, yet work within, naturalized acoustic norms that, when experiencing these productions, we are given a choice of how to respond to these situations.

In sum, then, *Songs of Ascension* and *Invisible Cities* poignantly offer a choice that is always already present. Given a sonic experience that could potentially expand our understanding of music, will we hear it nonnormatively as an opportunity to expand, or will we reject the challenge to our experience as wrong, closing our senses and minds to what is there? Do we accept sound (and our perception of it) in its (and our) specific (and sometimes unexpected) material, positional, and acoustical unfolding? Above all, we are not only confronted with these specific pieces in seemingly anomalous acoustic situations, but — at any acoustic moment — we are faced with a decision point: will I reify, and hence reinvest power in, the concept and perceptual schema of sound, rather than questioning it by negating an experience that did not adhere to the figure of sound? Or will I invest in the experience and realize that there is something other than what the figure of sound suggests? In other words, when confronted with these types of anomalous acoustic experiences, will I explore or reject them? And, ultimately, since the world — and our own relationship to it — is partly formed through the normative spatial-relational and acoustic mediation that we have internalized, do we explore or reject anomalous encounters with other human beings?

In the end, it comes down to developing an awareness of the seeming anomalies before we can even notice that we have a choice between accepting or rejecting what have been deemed nonnormative sonic experiences. Our awareness of that choice is the linchpin not only of this chapter, but also of the overall arguments of this book. In this chapter, then, we have examined some examples of the lived pedagogy and practice that lie at the heart of each such decision point. Essentially, I argue that what might look on the surface like an aesthetic appraisal is actually a choice made by a body trained in spatial-relational acoustics and encultured to orient itself to the figure of sound. Of course, this choice is crucial not because of its acoustic concreteness, but because it carries consequences for our overall relationship to the world. To put it another way, following the trails of experiential conundrums and decision points, *Songs of Ascension* and *Invisible Cities* offer vivid examples of the ways in which the world — and the figure of sound — is rendered through acoustic mediation, and the degree to which we have internalized this rendering. We will now move on to the third naturalized area of voice, listening, and to the music this book considers: sound itself.

MUSIC AS ACTION
Singing Happens before Sound

"Yes. I Can Hear My Echo": Vocal Paralysis and Vocal Ontology

In Richard Serra's 1974 video art piece, *Boomerang*, Nancy Holt speaks.[1] Simultaneously, she listens to the electronically mediated echo of her own voice. Between her utterance and its echo is but a slight delay.[2] In the bluish-tinted image, Holt wears headphones and, as though to get her bearings by touching something tangible, she holds each ear pad with her hands. Hearing her own voice consistently, predictably echoed—hence the title of the piece—she reacts spontaneously even as she reflects on the experience. She is charged with simultaneously producing and sharing her perceptions. Near the beginning of this ten-and-a-half minute piece, Holt says:

> Yes. I can hear my echo.
>   And the words are coming back on top of me.
>   Eh . . . [We can hear the thought process momentarily breaking down; Holt is seemingly unable to continue thinking while listening to herself. For me this is a key moment. She is moving into a nonlinguistic vocal space.]
>   The words are spilling out of my head and then returning into my ear . . .
>   It . . . puts a distance . . . between the words and their apprehension . . . or their comprehensions.

The words coming back . . . seem slow . . .

They don't seem to have the same forcefulness . . . as when I speak them.

I think it is also slowing me down . . .

I think . . . that it makes my thinking slower . . .

I have a double take on myself . . .

I am once removed from myself . . .

I . . . am thinking . . . and hearing . . . and filling up a vocal void.

I find . . . that I am having trouble making connections between thoughts . . . [There is a long pause—this is also a key moment.]

I think . . . that the words forming in my mind . . . are . . . somewhat detached . . . from my normal . . . thinking process.

I have a feeling . . . that I am not where I am . . .

I feel that this place . . . is removed from reality.

Although it is a reality already removed from the . . . normal reality[3]

Holt confirms that her monologue was not at all scripted but was a "totally spontaneous" reaction to the situation. This includes her use of the word *boomerang*, which Serra then adopted as the title of the video artwork.[4] She adds that this was the "first time [she] was exposed to the sound delay situation."[5]

Boomerang premiered less than a day after it was made—on the very night of its completion—before "a group of artists" at the Electronic Arts Intermix (EAI).[6] Founded in 1971, EAI was one of the first nonprofit organizations in the United States explicitly dedicated to video art, which was then a nascent art form. In analysis and criticism, *Boomerang* is, therefore, typically addressed within the context of video art. That is, the video images are the focus of discussion. For instance, it is in such an oculocentric context that Rosalind Krauss sees *Boomerang* as an exemplar of an aesthetics of narcissism. She describes *Boomerang* as "self-encapsulation," "a situation in which the action of the mirror-reflection (which is auditory in this case) severs her [Holt] from a sense of text: from the prior words she has spoken; from the way language connects her both to her past and to a world of objects."[7] Krauss echoes what Holt says during the performance: "I am surrounded by me." Thus, the piece, *Boomerang*, raises the question about presence by exposing how the medium of video trades in modes of reflectivity and simultaneity.

Two and a half decades later, Anne Wagner's analysis adds nuance to the insights about the video aspect of the piece. Instead of considering Serra's medium (video) alone, her discussion centers on the process documented by the video: Holt's public subjection to her own echoes. "We see her staring into a void, out of which language falls because technological artifice makes it too present, too insistent, too public, to be endured," Wagner observes. "Though the gap is simulated and correctable, its effects really happen; watching Holt struggle with a toxic media overdose, the viewer encounters something she can only be convinced is real."[8] In other words, as the video artist Laura Malacart writes, *Boomerang* is a "use of the voice that undermines semantic communication": the fall of language is consequent to the speaker's "affect and discomfort."[9] These reflections by Krauss, Wagner, and Malacart all rest on the same assumption: they presume a collapsed understanding of language as voice.[10] According to this shared understanding, which is evident even in Holt's self-reflections, her fluid ability to think of words is hampered by the delayed playback of her voice.

In contrast to the videocentric readings, the concrete video portion of the piece demonstrates a lack of confidence in audio. In my view, the video image of Holt struggling to conceptualize and verbalize her disorienting experience merely functions to confirm that the vocal disruption she is suffering takes place in real time and is not manipulated. It seems that the audio recording alone could have conveyed the point of the piece, yet Serra chose to include video as well. Why? One answer has to do with the evidentiary capabilities of sound and image. Serra's insistence on the visual image implies that audio alone is insufficient evidence of the trauma that results when one attempts to talk and listen to oneself at the same time. In that regard, Serra echoes the attitude to audible versus visible evidence emblematized by the term *hearsay*. Alternatively, as I will discuss later, sound in relation to the visual might be what Jacques Derrida would call the supplement.[11] In this paradigm, which prioritizes visual over sonic evidence, visual documentation accompanies the sound recording of the event to lend credibility to the artwork's overall veracity.[12]

At the heart of *Boomerang* lies the paradox that one's own words jam the ability to produce other words, interfering with articulative fluidity. Holt is subjected to the experience of the delayed effect of her very own narration of the experience of the effect that the delay has on her. As she describes her experience, she hears a slow recitation of that very description interspersed with pauses and nonlinguistic indicators of her own hesitation ("mmmm," "hmmm," "ehm"). The pauses are so frequent and drawn out that the boundaries between sentences are far from clear. Reflecting on this, Holt says: "Sometimes I find that I can't quite say a word because I hear the first part of it come back, and I forget the second part or my mind is stimulated in a new direction by the first half of the word." Holt's ability to speak is inhibited by a multiplicity of real-

FIGURE 3.1 · SpeechJammer (photo courtesy of Kazutaka Kurihara and Koji Tsukada).

time input, including the delayed playback of what she has just enunciated. That is, she is taken out of her train of thought by the sound of the sentence she is attempting to complete.

The experience of digital feedback in communication environments such as Skype is analogous and familiar to many of us.[13] In the event of a malfunctioning connection, you hear feedback and an echo of your own voice as you speak. If you listen to this echo, retaining a train of thought becomes challenging. Listening to one's own voice slightly out of time disturbs the thought process and its verbal articulation (speech). Considerable effort must be exerted to concentrate on what you are saying, rather than on the auditory evidence of what you have just said.

This cognitive challenge has been used to some advantage in the recently created SpeechJammer (figure 3.1). This "portable speech-jamming gun" is a device comprised of directional microphones and speakers and is intended to silence individuals who talk too much in a group situation. When the gun's user applies the trigger, the device records the "target's" speech with a directional microphone, firing his or her words back at him or her via a speaker. More precisely, in the words of Kaztuaka Kurihara and Koji Tsukada, who developed

SpeechJammer, "human speech is jammed by giving back to the speakers their own utterances at a delay of a few hundred milliseconds." This phenomenon, known as speech disturbance by delayed auditory feedback, is intended to cause the target to stutter and fall silent. The speaker feels "disturb[ed]" but does not experience any "physical discomfort," and the disturbance, the feeling of being "jammed"—the interruption of the thought process that causes the victim to stutter—disappears as soon as he stops speaking.[14]

SpeechJammer might appear to be a novelty device: it won an Ig Nobel Prize in 2012, an award "intended to celebrate the unusual [and] honor the imaginative."[15] Nonetheless, it challenged me to think about the relationship between language and voice, and it has inspired other creative projects around the world. Instructions for creating homemade SpeechJammers and speech-obstructing software applications (hereafter, apps) appear all over the Internet.[16] YouTube offers several examples of speech jamming in action, in situations ranging from social gatherings and video blogs to talk shows.[17] In many cases, speech-jammed victims dissolve helplessly into giggles.

Speech jamming works, causing victims to feel "zapped," as one app puts it, because, to many people, vocalization hinges on making sense.[18] A "jammed" voice, reduced to sounds like "hmm" and "ehm," fails to make linguistic sense. In this paradigm, then, voice and logos are equivalent; the voice functions within a closed system made up of known or knowable sounds. Utterances are judged in terms of their fidelity to a predetermined notion of what the sound should be, a notion determined a priori by the system. A victim of speech jamming can only make nonlinguistic sounds—sounds that fall outside the system. If, as Adriana Cavarero aptly observes, in "the logocentric tradition" the voice is defined as "words of a language in front of a mouth that opens," then nonlinguistic stutters and sounds of hesitation are not voice.[19] If anything, these sounds are heard as failed voice. However, in examining them, we might be led to think beyond abstractions of sound (and subsequent divisions into signified and signifier) to something that lies at the heart of all vocal modes, including those typically deemed nonsensical and risible.

The three examples I have touched on—*Boomerang*, the Skype malfunction, and SpeechJammer—turn on the paradox that throwing someone's own words back at him or her induces vocal incapacity or a sense of being blocked. We may observe various levels of breakdown, from Holt's articulate but halting description of how thought disturbance feels to incredulous stuttering and the collapse of the voice into self-deprecating laughter at its own lack of control. In each case, speech disintegrates from coherent phrases into nonlinguistic sounds and pregnant pauses. Holt describes the combined result of this peculiar speech as an absence of "forcefulness" in the words. As the speaker, she senses that she is "slowing" down or that she is "once removed" from her own speech. Furthermore, she feels that she is "thinking and hearing and filling up a vocal void."[20] Indeed, in all three examples, the vocalizer felt that she was genuinely impaired, relative to the linguistic system. What might engender such strong reactions to the breakdown of speech and exposure to other aspects of vocality? The unvoiced assumption is that voice functions only in the service of rational thought and speech.

But while the connection between thought, vocal intent, and speech is altered by delayed vocal feedback, to what extent is the voice truly paralyzed? In none of these cases do nonlinguistic utterances suffer any distortion or muting. *Boomerang*, for example, caused no impediment to Holt's ability to produce sounds like "ah" or "um." SpeechJammer's inventors observe that "speech jamming never occurs when meaningless sound sequences such as 'Ahhh' are uttered over a long time period"—the device jams only linguistic vocal sounds.[21] In fact, any notion that the voice is jammed or paralyzed by these processes hinges on the erroneous assumption that speech alone—that is, logos or sensible linguistic utterance—counts as vocalization. The experiences I have described caused breakdowns in vocalizers' ability to produce the vocables that they needed to pronounce correctly to be understood. What they experienced was not vocal paralysis, but a greater difficulty in their attempts to match their vocalizations to known sonic models—that is, to familiar words. Only preconceived sounds were jammed. The vocalizers' discomfort shows that, while there are plenty of vocal sounds involved in communication, we do not endow each sound with communicative value, though we do accord that value to sounds such as words: nonlinguistic sounds and pauses are understood as the words' negative backdrop.[22]

This is one example in which listening proves to be always already deeply encultured. Through a cultural process that divides signifying vocal sounds from nonsignifying vocal sounds, we learn to value each differently.[23] We learn to concentrate on vocalizations that reproduce signifying sounds to the extent that we naturalize them, and consequently we are unable even to conceive of other vocal sounds as vocalization. The examples discussed here expose the gaps in such naturalized listening practice and insist that voice cannot be defined by logos, or systems spun out of logocentrism, alone.

This book's previous chapters describe the naturalization of basic musical components. In the vocal experiences that comprise *Boomerang*, bad Skype connections, and SpeechJammer, the naturalization of yet another musical element, signifying sound, becomes evident. It is on this incomplete engagement with selected aspects of the thick musical event that our understanding of voice and music as well as our analytical tools are based. Consequently, our knowledge of voice and music is not only incomplete, but also skewed. It is not voice—vocalization as physical activity—that is stymied in these examples. Even when words refuse to come, the vocalizing body remains active. Instead of the logocentric definition of voice (with which Cavarero takes issue), *Boomerang* and SpeechJammer exhibit a voice that "transcends the plane of speech" and indeed "plays a subversive role with respect to the disciplining codes of language" and the fetishization of certain types of vocal sound.[24]

While the notion that music consists of selected signifying sounds has already been interrogated, I want to go one step further.[25] In the same way, as I discussed in chapter 1, air has been naturalized as sound's propagator, so that sound's passage through other materials—and its idiosyncratic nature in each unique material propagation—are not accounted for when we talk about music (except in extreme situations such as a thumping bass or a glass splintered by a piercing high note). And, as I discussed in chapter 2, while sound is always produced from a particular point in space, it is typically described as frontal, two-dimensional, and static when it is involved in music—again, unless there is an extreme situation at hand. I want to go so far as to posit that the third naturalized parameter of music is the notion that music's major identifying component is sound.

This chapter considers the ontology of voice, assuming a notion of voice that takes into account a number of activities related to sounds produced by living, communicating, and perceiving bodies. I will begin, though, by looking at how the marginalization of activity in favor of finished products affects the reception of visual artworks. In calling attention to this analogy, my goal is to show that just as the thick event of painting has been reduced to visual marks, the thick event of music has been reduced to sound.[26]

Considering Action: Jackson Pollock

Echoing the tension I have addressed regarding sounds and pedagogies of listening that legitimize classified sounds, Jackson Pollock's action painting unsettled long-standing positions on the dynamics at work in painting between causal actions and signifying or nonsignifying visible results. Thus Pollock's work effectively disrupted normative discourses surrounding visual art by provoking discussion about action in relation to its result, the painted marks.

Here, we observe an analogy to *Boomerang* and SpeechJammer. While a variety of vocal sounds were present, only selected sounds were considered within the value system of normative voice. Hence, the "umms" were seen to reflect the extent to which the normative voice's abilities were jammed. Analogously, the thick event of painting gives rise to a variety of marks, from drippings to be washed away (like "umms" to which we don't listen) to marks considered painterly (like words to which we pay attention). Pollock's explicit use of dripping signals his unabashed inclusion of elements normally seen as superfluous to painting: accidents resulting from lack of control, excesses to be covered or washed away. Pollock gave these marks value within a composition.

Pollock was explicit about the value he placed on the premeditated mark—that is, what made a painting look a certain way as a result of careful planning and the technical ability to carry out this plan. When a reporter dared to compare his work with aleatory art, the artist became enraged. He burst out: "Don't give me any of your fucking 'chance operations.'" To demonstrate that chance played no part in his process, he threw some paint at a doorknob across the room. Pollock "hit that doorknob smack-on with very little paint over the edges" and topped the gesture off with the comment: "And that's the way out."[27] However, while Pollock expressed outrage at suggestions of anything but traditional painterly values, and while he demonstrated his precise painting skills, his bravura created interest in the actions that lead to the existence of paint on a given surface.

Echoing Pollock, in discussing the 1950 *Lavender Mist*, Robert Hughes, *Time* magazine's art critic, also stressed the premeditated aspects of the compositions and the artist's technical ability. According to Hughes, Pollock's refinement as a painter results in "delicacy—at a scale that reproduction cannot suggest":

> It is what his imitators could never do, and why there are no successful Pollock forgeries: they all end up looking like vomit, or onyx, or spaghetti, whereas Pollock . . . had an almost preternatural control over the total effect of those skeins and receding depths of paint. In them, the light is always right. Nor are they absolutely spontaneous; he would often retouch the drip with a brush. So one is obliged to speak of Pollock in terms of perfected visual taste, analogous to natural pitch in music—a far cry indeed, [sic] from the familiar image of him as a violent expressionist.[28]

Generally speaking, although Pollock explicitly engaged in what was then a new process, discursive frameworks (even the artist's own) persistently relied on existing evaluative models, basing evaluations of his work on relationships to known indices. However, even if Hughes's defense of Pollock's work was cast in a language of precision and premeditation, Hughes's critique paved the way for action, rather than the final imprint, to be understood as the point of painting. If Pollock is admired for being able to hit a doorknob with paint, he is perhaps the first painter to be praised for the mere act of getting paint onto a surface. Therefore, despite his intentions, his work also challenged his viewers to expand their understanding of the types of marks that constitute a painting. Thus, it was not only the apparent free flow of Pollock's paint, between the two harbors of brush and canvas, that thrilled his audiences. In other words, even if it is in part Pollock's precise indices (a traditional painterly currency) that are admired today, it is because these indices were created by particular—seemingly unpainterly—actions that the resulting marks are celebrated. Hughes's praise of Pollock's "perfected visual taste" foregrounds the fact that all marks are preceded by actions.

Despite Pollock's conservative framing of his work, and his demonstrations of precisely placing paint on surfaces, he inadvertently presented a productive dissonance that was noted by critics and fellow artists. Harold Rosenberg, who is credited with coining the term "action painting," describes the shift: "At a certain moment the canvas began to appear to one American painter after another as an arena in which to act. . . . What was to go on canvas was not a picture but an event."[29] Rosenberg read Pollock's work as expanding the moment of art to include what happened before the arrival of the paint on the canvas. Rather than focusing on the result, in Rosenberg's view, Pollock stages his physical and visceral situation "'in' the [act of] painting" by laying his canvas on the floor, hovering over it with brush and paint, and allowing the final product to be whatever resulted from his movements. His focus was on what was happening: the canvas documented, and was part of, the event. Amelia Jones offers similar observations, noting that "the [Hans] Namuth images of Pollock show him standing above or within his huge canvas, overtly and theatrically *performing* the act of painting," and that in Namuth's 1950 movie, *Jackson Pollock*, the artist's "act of painting presented art as performance . . . rather than a fixed object."[30]

Going a step further, Jirō Yoshihara and the approximately twenty artists involved in Japan's Gutai ("Concrete") Art Association mistook Pollock's work for performance.[31] Inspired, the association created events (or happenings) and participatory environments such as moving in mud and making marks by leaping through a wall-sized piece of paper stretched over frames.[32] In an interesting twist, the group's idea of Pollock made its way to the United States, where artists began to undertake an intentional restructuring of painting as event. Today, despite the insistence by Pollock and some critics on the painterly mark, as Jones observes, his legacy is inseparable from the term *action*, and artists have found inspiration for performative work in the concept that term suggests.[33]

Considering Action, Again: *Noisy Clothes*

Working through this broad range of artistic practices suggests an expansion on Mladen Dolar's views: vocal sounds are not split into signifying and non-signifying, with the latter, by default, also classed as signifying because they bring the signifying sounds into relief.[34] Instead, in this chapter I suggest that, whether the result is audible to the ear through propagation in air or otherwise consciously or unconsciously sensed, a person who forms meaning based on a vocal utterance uses every aspect of that utterance in the process. Vocal utterances do not signify a static meaning in their capacity as a particular species of signifying sound but, rather, in their ability to cause a shift in a given person. And the entirety of that shift in a material and sensory relationship is used as the basis of meaning formation.

The Gutai Association's thought-provoking work offers one possible response to the following question: might it be possible to imagine a situation in which an imprint or a sound is not evaluated according to a preconceived value system?[35] What might result from the range of actions, such as those carried out by Pollock or artists associated with the Gutai Association? What kinds of results might follow musicians' actions? In my artistic practice, thinking about music making as action suggested a shift in its objectives away from signifying and preevaluated sounds. I looked to action in the hope of setting up a situation in which I could be driven by impulses and rewards other than the creation of particular, premeditated kinds of sounds — precisely because such premeditated sounds had previously imprisoned my music making within sonic ideals.[36]

The resulting musical experiment, *Noisy Clothes*, provided a physical framework for performers to enter.[37] It consisted of a staged setting, including instruments in the form of costumes.[38] Performers were given no knowledge of the instruments before they arrived on the stage — thus they were invited to explore the relationship between action and sound that resulted from their spontaneous interaction with the equipment.[39] The thirty costumes, created by Elodie Blanchard and me, were also sound-producing devices that would make sound only if the body engaged with them, launching them into action (figure 3.2). However, exactly how each costume produced sound depended on how each performer used it (figure 3.3). Unlike a conventional instrument such as the piano—which, as is commonly known and suggested by the keyboard interface, produces sound when fingers press its keys—our noisy clothes had no prescribed conventions or movements by which to create a sonic result (figure 3.4). Rosenberg's description of action painting provides a worthy analogy: "The painter no longer approached the easel with an image in his mind; he went up to it with material in his hand to do something to that other piece of material in front of him. The image would be the result of this encounter."[40] Similarly, music in *Noisy Clothes* resulted from the encounter between interacting, moving bodies and clothing.[41]

In other words, because of my experience with *Noisy Clothes*, I could ask: now that we understand the relationship between action and its result in the form of a range of marks or vocal sounds, how can we free ourselves from predetermining the range of marks and judging them according to a preexisting value system? The discourse about Pollock exemplifies the tension between these two positions and suggests that one cannot easily coexist with the other. It is only by altering the frame that we can set aside our inherent assessment of one set of judgments—that is, change the values attached to judging sounds according to a preexisting value system. In the case of Pollock, the former frame is the conventional painterly aesthetic value of the imprint's precision and overall composition. In contrast, the new frame within which Rosenberg and the Gutai Art Association considered Pollock's work rendered it more akin to an event.

In the case of a concert, the equivalent former frame is made up of the conventional music-based values that assume that sound is present and that music deals in sounds. In contrast—by involving designers, who deal in fashion, and inviting many performers who had no formal background in music to participate—*Noisy Clothes* entered a new frame that rendered the concert an event. Moving from a product frame to an action frame directs the participants' and audience's expectations, attention, judgment, and evaluation of a given event: what would be judged within one frame according to adherence to a preconceived sound system would, within another frame, not be held accountable to any sound system.

In response to this realization, Blanchard and I actively sought to reframe sound making by rethinking the collective rehearsal process. To discourage participants from forming expectations in relation to music making and from relying on the dominant belief that the project dealt in signifying sounds, we made a point of emphasizing that *Noisy Clothes* was a playful, social, experimental event (new frame) rather than a concert (conventional frame). To counter any notion that what we were doing should conform to the standards established

FIGURES 3.2–3.4 • Above: Three *Noisy Clothes* costumes (drawing by Juliette Bellocq; tracing by April Lee). Next page, top: Noisy clothes wearers' interaction (photo courtesy of the author). Next page, below: Silhouettes of a number of the *Noisy Clothes* costumes stored between rehearsals (photo by Elodie Blanchard).

by any particular musical culture, we invited nonmusicians to serve as performers. These decisions, which unsettled the traditional setting for musical practice, were intended to dislodge the judgmental mode of listening propagated by musical cultures, in which performed sounds are compared to standardized sounds, and to place performers beyond the reach of such judgments.

It is challenging to change habits. Nonetheless, *Noisy Clothes* was a success in relation to the experiment of freeing ourselves from the frame that solicits predetermined sound making; observations confirmed that the performers probably made their decisions independently of preconceived notions about sound and sound making. Because the instruments used in *Noisy Clothes* did not seem like instruments, we managed to shift the performers' frame of reference from playing an instrument to simply playing around. The boisterous play, laughter, and conversation that went on in rehearsals indicated that the performers were busy discovering rather than judging the sounds they made. For example, the group wearing the Velcro costumes hitched their arms and legs together in complex human-Velcro bundles, bursting into laughter as they disassembled the bundles and heard the result. This example of focusing on discovery showed performers operating outside of sound-focused framing, which begins to suggest answers to the question posed earlier: by focusing on action rather than the action's symptoms, we displaced the event from a preexisting value system. The parallel with painting is that, when the discourse about Pollock's work shifted to focus on action, his marks were no longer judged on the same basis as those featured on canvases that were not viewed as action paintings. Replacing one frame of understanding with another (even if only in a limited set of instances) opens a space in which to question the application of that frame more broadly. In other words, the power of a given frame has been denaturalized.

Conflating instruments with clothes and cross-fertilizing the performers' conceptions of each, *Noisy Clothes* opened a space that promoted play and exploration and, as a result, managed to counteract the tendency to operate according to an idea of a sonic outcome determined prior to any sonic creation. According to my observations during rehearsals, performers discovered sound after carrying out an action: sound was understood as a consequence of movement rather than as an attempt to match an a priori sound ideal, and performers were open to any possible result, sound being only one possibility. This revised conception enabled those involved to access an action-based process rather than a sound-focused music product.[42]

To summarize, through this performance-based research I realized that shifting frameworks and definitions may illustrate another side of the thick event. Applying one frame to Pollock's work, we understand him as making precise and premeditated marks that form a deliberate composition; employing another frame, we see the canvas as merely a documentation of the actions that took place. Thinking within one frame about an event, we understand it as a jump; using another frame, we understand the same event as a thumping sound. In a traditional musical context, the latter understanding, which limits our reading of the thick event to its sonic aspect, would serve as a list of desired outcomes for the goals of performance: our goal is to create a sound similar to that thumping sound. The former reading implies something more akin to choreography: our goal is to jump in this exact way. In a third reading, based on the listening pedagogy derived from *Noisy Clothes*, the primary focus is to jump for the sake of jumping, with only a secondary interest in discovering what sonic implications this action presents.

Detaching music making from conventional frameworks can provide an opportunity for the radical rewriting of prevalent notions of sound, listening, and action in relation to cause and effect. Such rethinking would in turn call for reconsiderations of what exactly is involved in sound and music making.[43] *Noisy Clothes* gives participants a way to escape the common musical dynamic ruled by a lurking, policing ear, which takes the form of the performer's own knowledge of predefined sounds and of her or his own attempts at sound making within sanctioned parameters. Released from preconceptions, we emerge from the acoustic shadows cast by our very own panopticons' watchtowers. It is this "automatic functioning of power," to quote Michel Foucault, and total autosurveillance that lead to self-regulation, which in turn recognizes and produces only recognized sounds.[44] The sound-making process itself becomes irrelevant and escapes consideration.

I want to suggest, therefore, that *Noisy Clothes* demonstrates the aesthetic value of the process of sound making, regardless of its final product, by designating any sound resulting from action as music. As in my considerations of *Boomerang* and SpeechJammer, we may conclude that the process is never exempt from sound; sound-making bodies are never irrelevant or paralyzed where sound is concerned. That is, aesthetic value is neither tethered to nor hinges on sonic results. Instead, *Noisy Clothes* points toward the line of argument I will develop further in the remaining chapters: any incidence of aestheticization or any other value judgment is contingent on specific material-relational dynamics. Often, it is naturalized parameters of music that mask these material-relational dynamics.

Engaging in music making through naturalized lenses can be counterproductive. First, as we witnessed with *Boomerang* and SpeechJammer, the regulating effect arising from the sonic taxonomy—normal versus pathological sounds—immediately causes us to become less efficient at producing the sound that we have been conditioned to value.[45] Production is mangled by self-surveillance. Second, because musicians and analysts tend to value music primarily that consists of such standardized sounds, other music—containing other kinds of sound—remains unaccounted for. Third, this selective evaluation prevents us from gaining access to music as a thick event. Music is thought to consist only of a particular sonic end product (not necessarily a given sound, but even just the presence of sound). However, in works like *Noisy Clothes*, corporeal action replaces standardized sound as the most important aspect of music. In this situation, music making becomes an activity that is not restricted by preconceived signifiers.

Body Music: A Chamber Opera without Vocal Cords

Adopting this sense of music making, we may begin to consider more than just gestures and activity located in a discernible area. While perceptions and descriptions of the voice have certainly been used to essentialize the body, as discussed in chapter 1, scholars who have thought through the anatomy of voice in dynamic relationship to repertoire may offer useful perspectives. Additionally, for example, Raymond Knapp and Mitchell Morris outline the inner anatomy of the voice and its relationship to vocal characteristics in specific tessituras, such as chest voice and head voice.[46] While their reading specifically traces sonic and stylistic characteristics of selected musical theater repertoire, beyond unlocking some of the specifics of the genre, it provides a model for reading vocal stylization through the singer's vocal apparatus and use of the body.[47] In a related view, David Sudnow offers in-depth reflections on how it might be the body—in his case as a pianist, the hands in particular—that leads music making. While in this example, the gestures of the hands, arms, shoulders and torso are visible, the pianist is still an interesting example for this discussion—first, because there are inner dynamics to which we as audience members are not necessarily privy, but which are key to the nuance of the musician's touch. And second, an important part of Sudnow's argument is that, through practice, the body gains knowledge that at times drives the artist, rather than the pianist commanding her or his hands to play in a given way.[48] Taking cues from these instructive works and others, my investigation and reading focus on areas involved in singing that are invisible to the naked eye. Building on the work of these scholars and others, I consider the body's movements as actions, and their central role in music making as it takes place through song.

In addition to the indiscernible nature of much of what makes up singing, the second central point of the project that I will discuss is the notion that the ontology of singing is masked by our fetishization of sound. That is, because the vocal cords produce such beautiful sounds, they traditionally get all the attention, misleadingly subsuming the multifaceted collection of events that comprises singing into sound alone. Contrastingly, if we define singing as action, singing can and does happen independently of the vocal cords.[49] The rippling layers of bodily activity that constitute singing may ultimately be filtered through the vocal cords; however, other mediators can also transduce and communicate the body's activities. Thus, moving considerations regarding singing beyond its various manifestations reveals that the singing body extends beyond that which we conventionally recognize as the vocal instrument.

The piece *Body Music*, the research phase of which forms the final case study I will work through in this chapter, takes to heart the notion that singing is an internal corporeal choreography.[50] On the basis of concepts derived from *Noisy Clothes*, *Body Music* makes music by composing actions with detailed attention to the internal, invisible choreography that yields vocal sounds. With this piece I began to build a vocal practice around a deliberate shift in attention from the vocal cords to the actions of the total body. Hence, taking to an extreme the premise that music ought to be defined as corporeal action rather than as sonic product, *Body Music* experiments with voice sans vocal cords.[51] (In this chapter I discuss the development of the vocal part only.)[52]

I asked the Miami-based Colombian composer Alba Fernanda Triana to develop *Body Music* in close collaboration with me. The project's development was experimental, experiential, and process-based. We began the experiments by identifying an inner corporeal vocabulary in biweekly workshops through the fall of 2007. After working independently on the material with only intermittent meetings, we resumed weekly workshops in 2011. I also worked with Pai Chou, an electrical engineer, and Luis Fernando Henao, a programmer and sound designer, to develop the necessary sensors and discover the range of possibilities available for mediating the data we would read from the vocal body. At the production stage, which we will enter in the end of 2015, additional collaborators who have only been involved in the discussion stage thus far — a digital visual artist, a fashion designer, and a dramaturge — will begin to participate more actively.

To create the composition we mapped, analyzed, and finally expanded on the movements and internal activities that engage the singing body. This process was divided into three distinct phases. During the first phase, we observed conventional singing and mapped the activity that flowed into it. In the second phase, we organized and expanded on the activities we had mapped. Finally, during the third phase, we created the composition with these activities and general processes as starting points. When anyone asks me, I describe the vocal method and its mapping as modifying breath and the breathing process through the study of all aspects involved.

The result of the first phase was to increase our general understanding of the bodily actions and materialities that influence sound. We found that these were divided into voluntary and involuntary processes or activities, and that both groups potentially affect vocal output. Involuntary processes, or physical changes in response to other activities, include changes in the body's heart rate, respiratory rate, sweat rate, and hormonal level.[53] Our position echoes Tia DeNora's: "To the extent that music and body are linked, music's properties may come to anchor situations of action. It [sic] may do this by anchoring embodied (and by no means necessarily conscious) practice, including physiological features such as pace, energy, comportment, skin tone, and arousal levels (muscle tone, heart rate, breathing, perspiration, endocrine function)."[54]

We took into consideration the fact that, while we cannot directly govern the heartbeat, overall bodily activity—including breathing and singing—does indirectly influence the heart's tempo. Furthermore, while singers in training are not necessarily trained to manage involuntary processes, the transformational power of those processes over the body has the capacity to influence overall vocal output. Therefore, while we did not have direct control over the body's involuntary activities, because they would influence traditional vocal output (vocal fold signals) we wanted to consider reading them with sensors (in the project's final stage) and using that information as the basis for sonic and/or visual output. In doing so, we hope to show that the vocal cords are only one possible interface that sonorizes the overall bodily activity of singing and the corporeal changes it causes.

The main process we worked with and through was breathing. Situated at the center of singing, breathing is a curious case of combined voluntary and involuntary actions and processes. The voluntary actions and processes related to singing center around breathing and movements that expel air out of the body. More specifically, these voluntary actions are limited to the particularities of inhalation and exhalation, such as tempo and intensity. However, at some point involuntary processes and breath command will kick in, creating a limit for slowness or rapidity as well as intensity. For us, then, involuntary processes and activities came to include the basic and constant cycles of inhalation and exhalation. (In all singing, the characteristics of inhalations and exhalations depend on bodily manipulations and actions.)

During the mapping phase, we began by observing and analyzing bodily activity, deriving from it a breakdown of the minute actions that together comprise singing. These catalogued actions were divided into two categories: (1) actions that move air through the body (in this case, we explored the nuances of inhalation and exhalation, including external forces that could influence the process); and (2) actions that shape the inner membranes and cavities through which air passes, generating various sonic timbres by varying the tautness of the flesh, skin, and membranes. Figure 3.5 presents a summary of the observations we made during our first series of experiments, in which we studied how various physical gestures affected the flow of air through the body. We noted the type of sound that could result from each movement (short or long sounds, for example) and whether or not the sound unit could be easily replicated.

Following the mapping and analysis stage, some of the experiments described in figure 3.5 show our expansion of the processes we first noted. For example, we investigated the many ways in which one may expel air held by the lungs from the body. We explored every movement imaginable that would affect the shape of the lungs. Remaining at rest is one example. Due to overpressure in the lungs, air will automatically escape after an inhalation if nothing is done to prevent it. Another example involves drawing in the stomach, an action that pushes on the internal organs which in turn push on the lungs, causing them to deflate. The resulting sounds ranged from something akin to a violent expulsion of air to something that might recall a sigh.

We also experimented with altering the shape of the cavities through which air passes: essentially the chest, throat, and nasal and oral spaces. Related to these is the energy the body (mass) expends while remaining at rest. The greater the tension in the skin and flesh, the denser the body's overall mass, which when activated will thus produce higher sonic overtones.[55] Like the head of a drum, facial skin and flesh produce a pitch when one taps them with a finger; again, like that of a drum, the pitch grows higher as the skin becomes tauter. For example, this is in part what allows us to distinguish the voice of someone who is smiling from that of someone who is not. Hence, not only the physical gesture of the smile, but also the energy of the given body mass at rest results in a particular overtone collection. We can also evoke the subtle but stark distinction between two different energy levels if we think about the difference between the vocal sound produced by a person with a spontaneous smile versus a forced smile, with a raised larynx versus a lowered larynx, and at rest versus in balance or during a fall.

Because the overall position of the body's frame makes possible but also limits overall inhalation and exhalation, we experimented thoroughly with the

Formula: A+B=C1+C2

	(A) Air Shapers	(C1) Type Sound		
		Short (one impulse)	ContRep	Cont.
1	Shoulders (arm/hands; torso; neck)			
1.1	Back/Forth [pretty even] (arm;	✓	✓	
	Tremolo			✓
1.2	Up/Down [not so even]	✓	✓	
1.3	Circles B/M [not so even]	✓	✓	
2	Arms (shoulders, torso, hands)			
2.1	Punching Front/Back	✓	✓	
2.2	Punching Sideways	✓	✓	
	(Punching Short Forward/Up works good)	✓	✓	
2.3	Punching continuously unidirectional punch (sound becomes bigger)		✓	
3	Legs (torso, hips)			
3.1	Kick Forward/Backward	✓		
3.2	Combined with arms (kick back same arm forward works good)	✓		
4	Abdomen (torso, neck)			
4.1	Panting (all dynamic ranges)			
	In (x1-)(repetition limited by too much air)	✓	✓	
	Out (x 1-<)(might be maintained longer due to little inhalations)	✓	✓	
	In-out patterns	✓	✓	
4.2	Inhale (short/long)	✓		✓
4.3	Exhale (short/long)	✓		✓
5	General body			
5.1	Full-body fall (release legs)	✓		
5.2	Step (could include inhalation; arms; push away with other leg)	✓		
5.3	DELETED	✓		
5.4	Jump (inhale, whole body energy, going downwards before going up; exhale, contraction/release)	✓		
5.5	Running (torso, arms, neck; contraction/release; breathing is affected)			✓
5.6	Shaking (whole body)			✓
5.7	Torso moving continuously (less defined; small dynamic range)(arms, neck, head)			✓
5.8	Tremolo	✓	✓	✓
	(B) Filters	(C2) Resulting Sound		
6	Throat (various degrees of openness and tension)			
7	Mouth (throat, jaw)			
7.1	Closed (various degrees of closeness)	✓		✓
7.2	Open			
	Vowels i - e - a - o - u	✓		✓
	Consonants			
	Respiratory s - f - h	✓		✓
	Short, explosive k - p - t	✓		
	Voiced b - d - g	✓		
	j - l - m - n - v - z	✓		✓
7.3	Nose opened/closed	✓		✓
8	Nose (various degrees of openness)(throat; mouth)	✓		✓

FIGURE 3.5 · List of early systematic work on the movements of *Body Music* (courtesy of the author and Alba Triana).

body's stance. Common ideas about singing relate this category to posture, but our exercises took this idea to an extreme. Any position—including a tightly rolled ball, being at rest, falling, and an expanded posture with limbs and spine stretched out—affects the body in ways that, in turn, affect its sonic range. For example, stance determines the ease or difficulty with which air enters the body. An expanded body allows air to enter easily, while a compressed body requires the movement of body parts and organs to make room for the air. If the body is in a compressed state, when air enters the lungs, which are themselves compressed by the surrounding organs, those organs must move to make room for the lungs to expand. Imagine an empty balloon buried in popcorn. When the balloon is filled with air and therefore grows in size, the popcorn will be pushed away to make room for the balloon's expanded circumference. In other words, while we can offer instructions about breathing, they are limited to the body's frame and posture, which in turn regulates tempo, intensity, and so on.

During our initial mapping and analysis of the various parts involved in and affecting breathing, we played with and, through experimentation, expanded the possibilities for variation in the basic principles. We were most interested in exploring how this continuous alteration could be interrupted or affected. Again, the principle on which we operated was that, while breath is the foundation of song, breathing arises first and foremost from a bodily necessity: the need for air, the vacuum that results when lungs are short of breath, and the physical compulsion to inhale.

Hence, we were building on breath's behavior as a bodily necessity, prior to being altered by the aesthetic demands of song. To this end we worked on various kinds of breathing, first considering what we referred to as the regular breathing pendulum. This is our term for regular breathing that calls no attention to itself—that is, the breath that takes place with regularity even when we do not will ourselves to breathe in a particular manner. We then looked to the extremes of the pendulum, to breathing that is out of the ordinary—in other words, to breathing modes that interrupt ordinary, subconscious breathing. These include deep inhalation, sudden inhalation, and inhalation through pursed lips. We developed a host of such interruptions and observed the exhalation that followed each kind of inhalation. These interruptions included a number of bodily manipulations and actions such as punching the air, kicking, thrusting the torso or entire body forward, pursing the lips, and closing the mouth—all which force air out of the body in a certain way or affect the character of the air flow.

The result is a piece in which the audience sees a performer who inhales and exhales in a variety of extreme ways, carrying out a choreography of sorts that changes the shape and impact of the torso, neck, and facial areas, all to engage and affect the vocal apparatus sans vocal cords.[56] Whatever sounds may result from this choreography are extremely intimate: timbrally rich, yet subtle and quiet.

To emphasize what the results of *Body Music*'s inner choreography sound like, though, would be to miss the point of the project: its attempt to reconsider music not in terms of sonic indices, but as the actions that comprise a thick event. Vocal cords' beautiful sound has the capacity to seduce us, but focusing our attention on that sound instead of on the action that produced it, when a notational scheme for that sound gains value, the sound or signifier is similarly reinforced. As I will discuss in chapter 4, when we approach singing as action rather than sound, we free ourselves from many inhibitions formed during our efforts to produce the correct sound, and we produce most sounds with much less effort. As Triana and I discovered, initial attempts to notate *Body Music* exemplified this tension.

This historically close connection between notation and sound caused unexpected problems during the process of notating the piece. In the third developmental phase, composing the piece, there was a period of approximately half a year when we were in constant negotiation over the notation and the way in which the compositional material seemed to transform once it was set in notation and performed from notated instructions. After months of failing to see why we were unable to agree on a notational system, we finally understood that the notational system we knew was unavoidably tied to sound. Thus, the series of divergences in opinions ultimately led to insights into the relationships among voice, action, and notation and how their (hierarchical) relationships are actually mediated by the overarching framing (a topic I will return to in the next section of this chapter).

The notational problem was instantly solved once we understood that when we put *Body Music*'s components into traditional notation and the performer accessed *Body Music* through that notation, he or she reverted back to performance of sound instead of focusing on the actions (see figure 3.6). It is important to note that this dynamic did not arise as a result of notation per se, but rather served as evidence of the values Western notation is asked to address, values that lead to a focus on sonic traces.

After a subsequent period of creating and testing a number of different *Body Music* scores, Triana and I were finally satisfied with a notational system we felt did not treat sonic results as its focal point. While the previous notation system we had worked with conveyed physical instructions and information that was instructions both for actions *and* for sonic results (dynamics, rests,

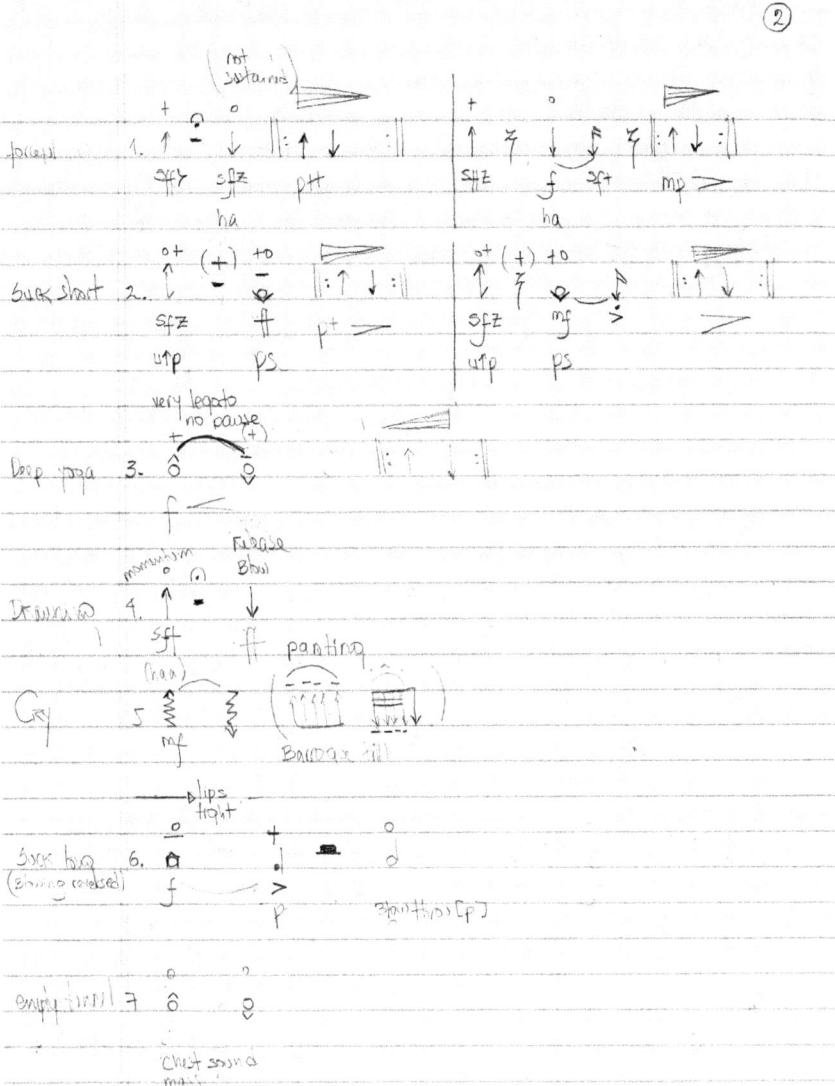

FIGURE 3.6 · A version of the score for *Body Music* from September 25, 2012. The phrase "Suck short 2" refers to both the fact that this is the second idea of the section, and the manner in which to inhale—with pursed lips. Upward pointing arrows indicate inhalations; downward pointing arrows indicate exhalations. The use of different lengths for the arrows is adapted from the Western notational system for music. That is, at the "Suck short 2" system, the first arrow (with a "flag") is shorter in duration than the third arrow (without a "flag") and the second arrow (with a tiny open circle and the plus sign+ refer to open or closed mouth, respectively). The syllables at the bottom refer to the approximate sounds produced (courtesy of Alba Triana).

consonants, and vowels), the new score's instructions focus on actions *rather than* the sonic result. The excerpt of the score in figure 3.7 shows inhalations and the actions undertaken while inhaling and exhaling. Thus, rather than the absence of sound, the rests refer to pauses in inhalation and exhalation. The second note (close to a traditional half-note) is about twice as long as the third (close to a traditional quarter-note); the whole-note pauses are approximately four times as long as the quarter-note, and so on (notes indicate relative durations, rather than specific sounds). The arrow and the words *back punch* indicate vigorous backward-thrusting arm movements. Finally, the written words such as *pendulum, etc.* indicate relaxed breathing. These are just a few examples of the ways in which the notation specified actions, yet did not indicate a desired sonorous outcome. Thus, unlike the particular vocal quality and timbral profile that traditionally define the operatic voice, *Body Music* is not defined by a given set of sonorous material.

Body Music's operatic plot revolves around the questions: What is voice? And how does the body voice? Indeed, the project takes seriously the notion of *Gesamtkunstwerk* and the gathering together of multiple sensory experiences within a single form.[57] And what audiences hear is the sound of the shifting shapes of chest, throat, and mouth and nose cavities and the differences in the speed of the air moving through them. These mechanisms and procedures are the foundations of conventional singing. Unlike conventional listening practices — in which sounds understood to be signifying are front and center, and nonnormative sounds are considered erroneous — *Body Music* attempts to avoid allowing the sounds of the vocal cords to overshadow those produced by actions of the internal organs. That is, with *Body Music* Triana and I aim to emphasize that the human body is always engaged in a vast variety of articulations, some of which happen to result in what audiences have traditionally considered as normative vocal sounds. In this way, *Body Music* sets out to avoid anything that can be used as a sonic signifier, and therefore it focuses only on action. It does so by exploring vocabulary on a microphysiological level, demonstrating that singing, which is normally understood as a form of sound produced by the vocal cords alone, is made up of several kinds of activity.[58]

By decoupling singing from vocal cords, Triana's and my experimentation had led to observations and a greater understanding of the process of singing: singing is action, as our difficulties about the notation showed us. In an interesting twist, the attempted documentation and notation of the piece wiped out those observations. However, the important point here is that it is not notation per se that erased the observations about singing. Rather, it is because the dominant notational system focuses on sound, and therefore the notational

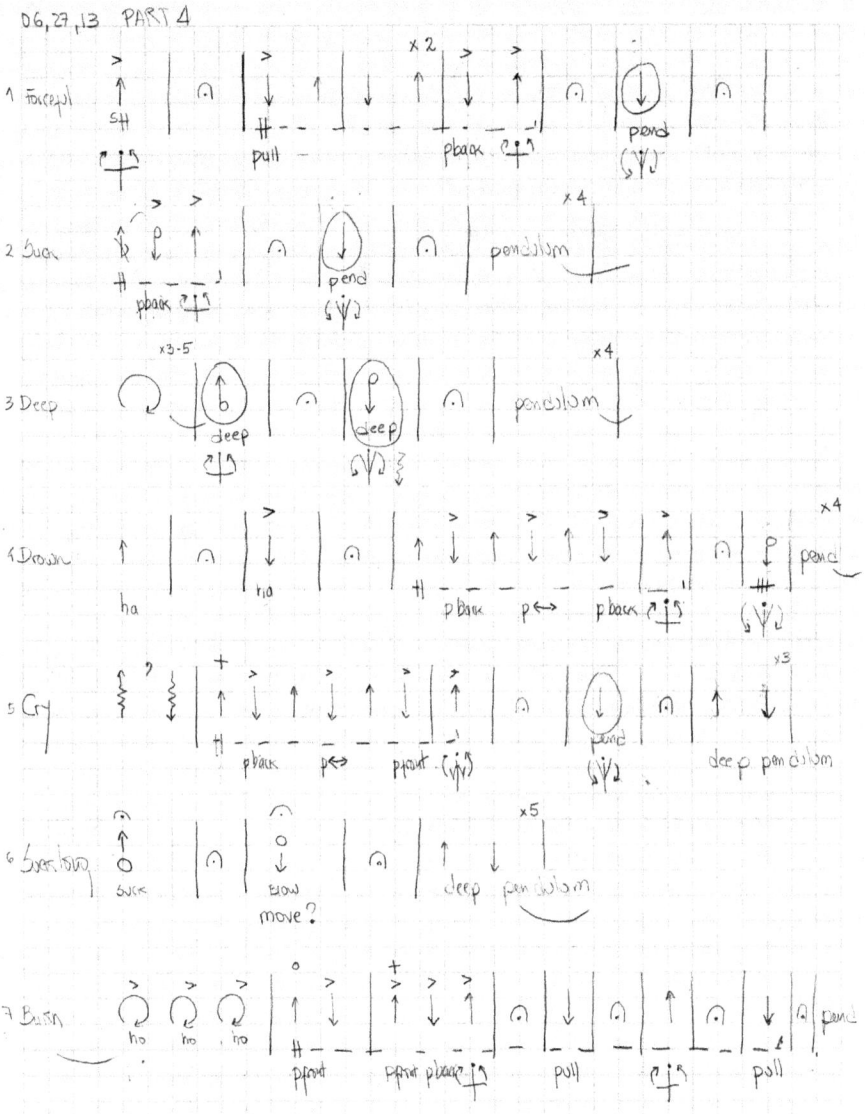

FIGURE 3.7 · A draft from June 26, 2013, of a section of the final version of the *Body Music* score. As in figure 3.6, the phrase "2 Suck" indicates that this is the second idea of the section and the manner in which to inhale (with pursed lips). The upward and downward pointing arrows and the length of arrows here have the same meaning as in figure 3.6. Stick figures indicate arm movements. The letter *b* indicates a punch-like movement backward, moving the arm and fist at the side of the torso and behind the back. Other abbreviations are also movement indicators (courtesy of Alba Triana).

system available to performers and audiences alike allows only sonic dimensions to be expressed.

Any dominant notational system is really just the actualization of dominant discourse. Furthermore, while notational systems allow for and promote a certain mode of thought and reality, sound does not have the capacity to express and think about other dimensions. If performers, composers, and audiences think about this in terms of Clifford Geertz's image of the thick event,[59] dominant discourse and supporting notation are limited to a few strands of the thick event, while—by necessity—others are not on the observer's radar. This limited selection is then reproduced as reality. To be sure, it is not the notational system per se that produces this limited reality, but the reality felt by singers and other vocalizers is that sound is also a naturalized parameter of voice and music.

How Experience and Meaning Making Are Limited

Musical notation's entry into the compositional process seems to be tethered to working with music through signifiers. When determining distinct units for notation, and when employing notation to contain a musical event, the thick corporeal event is necessarily subject to a reduction. And, as a result, our relationship to the event shifts after we access it via notation. Even in my attempt to work creatively with action to illuminate singing as a corporeal process, the process of notation unequivocally reduced the corporeal event to a relationship between sign and signified. Once sound, merely an extracted component of the thick event of action, was fixed in notation, a reduced and partial aspect of the thick event was frozen. Signification and related notation constitute just one example of the reduction of singing and listening. In the same way that the question about the falling tree is symptomatic of a tendency to ignore—or an inability to comprehend—a thick event, the process of arriving at notation for an event, it seems, threads the experience through the needle's eye of an a priori idea of sound: the part of the thick event that prior values and priorities have rendered notatable. Furthermore, the practical uses of the document tend to perpetuate these reductions. Of prime consideration in the process is the decision about which concepts to connect to the signifier, and how our relationship to that broader event shifts after we access it via the notation of the signifier.

Considering notation in relation to well-known theories of *phōnē* (the voice) inscription, and body, throughout the creation process of *Body Music* we can learn that the dynamic between voice and notation arises from an overarching paradigm. That is, this dynamic does not reflect inherent relative virtues of the formats of speech or writing, but rather the overarching value system that makes such discussions potent.

The dilemma that Triana and I experienced in *Body Music*, triggered by our conversation about notation, is representative of an ongoing debate. What is the possibility of reporting on the fine details of the thick event and rendering it into meaning? For Ferdinand de Saussure, this process requires a linguistic system, which he argues is most accurately embodied through phonemes; for Derrida, speech and writing are interconnected. With a slight twist, Roman Jakobson concentrates on the arbitrariness of the relation between vocal speech sounds and meaning-making sound units. Let us consider these influential theories below, in relation to the reduction I felt when notation was introduced in *Body Music*.

In broad strokes, Saussure and Derrida disagree on whether the arbitrariness and "unmotivated institutions" of signs deny evidence of any natural attachment between signified and signifier.[60] If the sign does not arise from any foundational reference to reality, does that mean that no one sign system (for example, speech or orality) is more natural than another (for example, writing)? Saussure suggests that sounds are related more intimately to thoughts than, for example, to the written word. Derrida's critique of Saussure, and Western philosophy in general, was predicated on a particular critique of and skepticism about phonocentrism—the privileging and romanticizing of language's acoustic dimension at the expense of the written. At the base of Saussure's system is the idea of a natural attachment or "natural bond" with sound.[61] That is, grounding language in the body through *phōnē* depends on the assumption that language and experience have an organic—to paraphrase classicist Shane Butler's vocabulary—and privileged relationship to one another.[62]

More specifically, Saussure posits that, prior to the linguistic system, sounds and ideas were not connected. In other words, before the linguistic system there was no way to evaluate baby babble containing syllables akin to *mama* and *papa*. The overall value system that vocal sounds should signal linguistic meaning—rather than a supposed meaning inherent to each syllable—engenders a pedagogy of listening and the kinds of distinctions that are detectable in listening.[63] "In a language," Saussure writes, "there are only differences. Even more important, a difference generally implies positive terms between which the difference is set up; but in language there are only differences without positive terms."[64] Outside a system of contrasting and related sounds—that is, outside the linguistic system—a sound's phonic substance would not mean anything. The introduction of /e/, /i/, /o/, and /u/ makes /a/ distinct as /a/ because of its distinction from them—hence /a/ becomes /a/ in its negative relation to /e/, /i/, /o/, and /u/. Furthermore, /a/ does not receive meaning unless it is drawn into further difference relationships—for example, through additional vowels and consonants in such combinations as *alpha*, *art*, and *aluminum*. Saussure believed this to be a general principle at the base of all linguistic signs: "A linguistic system is a series of differences of sound combined with a series of differences of ideas."[65] The distinctiveness of words, and of the concepts toward which they point, is grasped in independent and negative relationships of difference rather than in conceived, inherent, affirmative relationships that spring from norms within a linguistic system.

While Saussure believes that language's oral tradition is independent of writing (and that this independence makes the science of speech possible), Derrida is in complete opposition to this idea. Derrida argues that what can be claimed of writing—that it is derivative and merely refers to other signs—is equally true of speech. In the end, Derrida's critique is that Saussure made linguistics "the regulatory model, the 'pattern' for a general semiology," and that Saussure "for essential, and essentially metaphysical, reasons had to privilege speech, and everything that links the sign to *phōnē*."[66] For Derrida, this is a weak assumption. He claims that there are differences in writing that are not detected in speech, and thus the notion that speech and writing are separate is illusory. He believes that writing has been deemed an afterthought by most philosophers to keep their metaphysics intact. Furthermore, Derrida is far from certain that differences actually do exist in the world. If they do not, then writing is a system only, one spun out of conceived differences.

Derrida addresses this question through the *différance/différence* pair, where the only difference is between two vowels, and where written difference is not detectable through speech. He posits that writing is not secondary to orality or experience, but rather at the center of them. That is, words like *sweet* and *sweat* are not learned by attaching them to concepts and things. It is by comparing them to other words—by linking language to language (rather than language to reality)—that we learn to distinguish one from another. *Différence/différance* is an example of speech's reliance on writing. Furthermore, the pair's general meaning of *deferral* also refers to the process Derrida identifies: no fixed meaning outside a relational system, because meaning is always already deferred until words are related to one another (the meaning of *hot* is deferred until it is in relationship with *cold*).[67]

While Saussure and Derrida seek to untangle which format has more fidelity in relation to an assumed or imagined reality, Jakobson introduces a perspective that includes consideration of physical action in relation to identification and notation of a sound unit. Furthermore, while areas in which difference could be detected formed the jumping-off point for both Saussure and Derrida, Jakobson focuses on exposing the arbitrariness between sign and signified through the human anatomy. Jakobson argues that the phonemes that comprise the words *mama* and *papa* are sounds that every young child emits as a result of the anatomical relationship between the vocal and respiratory mechanisms.[68] When we breathe, air flows in and out of the air tract as well as the mouth. A baby must open its mouth in order to breathe, and it is this very movement that produces the phoneme /ma/ or /pa/. But keeping the mouth open at all times, a position that would yield a vowel sound, would dry out the oral tract. A gesture as simple as closing the lips serves, in part, to keep it moisturized.

Thus the phoneme combinations and signifiers *mama* and *papa* can be conceived not only as words, but also as the voicing of experimentation, play, and the mechanical functions of the body. Jakobson also identifies these sounds in the nasal murmur that often accompanies nursing. He writes: "The socialized and conventionalized lexical coinages of this baby talk, known under the name of nursery forms, are *deliberately adapted to the infant's phonemic pattern* and to the usual make-up of his early words."[69] Hence the presence of such consonant-vowel combinations—in all, 1,072 synonyms for *mother* and *father*—are found in nursery rhymes around the world.

In short, because Jakobson shows that the selection of audible sounds from a thick event, and the subsequent connection of those phonemes with the concepts *mother* and *father*, are effected by a listener who yearns for his baby to signify these ideas in the way he has already signified himself in his relationship to the child, Jakobson elucidates directionality in the process of signification. While the baby is in fact exploring the physicality of the vocal and respiratory apparatus, the father listens intently, taking slivers of the babble to be his interpellation.[70] The relationships among the vocal sound, the sign (in the form of selected sound segments such as *papa*), the inscription, and the signified are arbitrary. The father himself chooses the sounds /pa pa/ to mean the signified—him—for whom he would like the child to call, and for whom the baby indeed calls with its whole helpless and needy being.

In the same way that the corporeal actions of *Body Music* happened to yield a number of small sounds, some of which were selected, groomed, refined, and chosen for repetition, what we see as the potent words *mama* and *papa* may be viewed as sounds chosen from a sonic field made up of sounds and silences and granted entry to the linguistic realm. In other words, these sounds are chosen to be endowed with linguistic meaning, which moves our ears to no longer define them as babble but as communicative language. In this process, sounds that result from corporeal necessities and functions are invested with meaning and communicative value. Yet it is precisely because selected sounds are thus invested that vocal performance in *Boomerang* and SpeechJammer is thought to be impaired. Each of these theories arises as a result of our interpellation and endowment of a sound or inscription with meaning.

The impulse to select and endow certain sounds with meaning is not random, nor are the selections in question. The impulse and the effort to discriminate and prioritize arise not only from our evolutionary biology—we have needed to pay attention to the human voice over other sounds—but also because, as Butler summed the situation up to me, "people are trained to reify voice as sound."[71] The most basic value system is the favoring of sounds over the physical events that give rise to them. Secondarily, but more often considered and as exemplified by *Boomerang* and SpeechJammer, is the question of the types of concepts that are felt to be significant enough within that value system to be signified and concretized through the attachment of a signifier. In this instance, sound's importance is evaluated according to its strong or weak relationship to a system of signification. And in this model, the production of sound's meaning is assumed to be located in its hearing (or in the relation between speech and hearing).

But there is another way of thinking about signification in relation to the body's actions, and about *phōnē* in relation to inscription, that does not rely on fixed, a priori positions or on mutually exclusive categories of originals and copies.[72] For example, when the word *mama* is mouthed and the mouthing and breath sonorized, it is not only about the word *mama* in relation to the word *papa*, in turn in relation to yet more words. Rather, the physical unfolding of those words—or even a timbral modulation in their pronunciation, or even a timbral modulation that is not attached to a recognizable word—causes physical changes in the speaker or vocalizer, and it is from the sensation of that changed corporeal environment that we build meaning. In turn, as the corporeal circumstance inadvertently leads to sonic vocal events, the corporeal environment can also be affected by enunciations, such as "mama," "papa," "cold," "room temperature," or "man." Speech is therefore not arbitrary in relation to meaning making and reality, but neither does it unfold through a casual and nonmaterial chain of relationships between concepts, as outlined by Saussure and Derrida. Needless to say, this diverges radically from the theory that systems of meaning are based on our ability to recognize the differences between /a/, /e/, and /i/.

In relation to *Body Music*, as soon as inner activity was put into notation it felt as though the piece had been pulled into a different realm. Can we fully understand this tension by analyzing the situation through the relationship between voice and notation? Were we to do so in this specific example, the implication would be that notation had killed voice. While this was the conclusion to which I first jumped, informing my first conversations with Triana, I later came to understand that the tension between body, vocal sound, and notation is not inherent to their relationship. Instead it is an overarching framework that pulls them into a relationship of mutual exclusivity. Difference, then, is defined not only in terms of distinguishing, say, *a* from *e* (as in *pat* and *pet*), but also according to the overarching principles on which the system relies. For example, the designation of sounds as signifying or nonsignifying, or as defined in a negative register in relation to a signified—such as the mispronunciation of a word, a pitch that is out of tune, or what is deemed to be babble—is possible only when we buy into the overall premise of a system of difference.

Because medieval music represents an early stage of Western notation in which tensions between words, notation, sounds, and voices were regularly and explicitly taken into account, its study has proved instructive in considering the complex relationships between sign and signified, notation and performance, and sound and sense. For example, while early notational systems, such as the nondiastematic systems discussed by Leo Treitler,[73] were the least prescriptive among the many systems in operation in the Middle Ages, early chant notation was complex and multidimensional. Within this notation and performance practice, the concept of melody was inseparable from the human voice that produced it. In other words, the notational system represented voice.[74] In these early negotiations between performers, composers, and notation, the mismatched partnership between notatable sound, the senses, and meaning tells of the power of song and music and of how sound, performance practice, and notation can coexist. Importantly, this stage in Western notation shows a relational dynamic that is not premised on a strict notion of fidelity—either fidelity between the notation and the musical rendition or that between the composition and the performance.

Emma Dillon suggests that the effects of performers on listeners cannot be fully grasped through the study of grammar and musical systems alone.[75] She uses the term *supermusicality* to capture the sound of the singing voice and the work it accomplishes through that sound, which is unique and independent of the composition. It is only by "restoring a singing voice to these texts [that] an uncanny transformation of meaning occurs."[76] And it is here that debates about a stable meaning for a sound or inscription implode; here conventional meaning-making paradigms do not suffice. For Dillon, Augustine's *Confessions* and *Enarrationes in Psalmos* are instructive for understanding the medieval "anxiety about what singing did to the sound and sense of words"—for example, how a text about chastity could be delivered by a voice that listeners found sensually irresistible.[77] Indeed, this is one of Dillon's key examples of supermusicality—that unnamable aspect of voice or music, the attraction and power of which lie beyond the reach of understanding through fidelity to words or music. Augustine's dilemma "establishes a standard for musical sound in relation to verbal sound and meaning; but when resituated in the larger mediaeval discourse of words, it reminds us of the high ethical stakes of effecting a rift between sense and sound."[78] Augustine wrote:

> The pleasures of the ear had a more tenacious hold on me, and had subjugated me; but you [Christ] set me free and liberated me. As things now stand, I confess that I have some sense of restful contentment in sounds whose soul is your words, when they are sung by a pleasant and well-trained voice. Not that I am riveted by them, for I can rise up and go when I wish . . . but my physical delight which has to be checked from enervating the mind, often deceives me when the perception of the senses is unaccompanied by reason, and is not patiently content to be in a subordinate place. It tries to be first and to be in the leading role, though it deserves to be allowed only as secondary to reason.
>
> Nevertheless, when I remember the tears which I poured out at the time when I was first recovering my faith, and that now I am moved not by the chant but by the words being sung, when they are sung with a clear voice and entirely appropriate modulation, then again I recognize the great utility of music in worship. Thus I fluctuate between the danger of pleasure [in the music] and the experience of the beneficent effect [of the words], and I am more led to put forward the opinion (not as an irrevocable view) that the custom of singing in Church is to be approved, so that through the delight of the ear the weaker mind may rise up towards the devotion of worship. Yet when it happens to me that the music moves me more than the subject [meaning or truth] of the song, I confess myself to commit a sin deserving punishment, and then I would prefer not to have heard the singer.[79]

For Dillon and Bruce Holsinger, Augustine's account of listening, which vacillates between the linguistic and the "innately non- or even prelinguistic in music's flow through the human body,"[80] serves as a launching point for further investigation into the tension Augustine articulates between "words and their sound, and music's particular ability to complicate the sound-sense relationship, which clearly has roots in a broader linguistic theory."[81] For Dillon this exemplary point of tension serves as a poignant illustration of her sense that music has the "capacity to unsettle words."[82]

I understand Holsinger's interpretation of Augustine slightly differently from the way Dillon understands it. For Holsinger, the "pleasure of the ear" (*voluptates aurium*) is not dynamically pitted against "truth" or "meaning" (*cantus, quam res*). Rather, in Holsinger's words, "the human body represents . . . the very ground of musical experience."[83] Rather than being complicated by the flesh, "musical sonorities" are indeed "practices of the flesh."[84] Indeed, for Holsinger, music at its root is not divided into sense versus pleasure. It is only value systems (as personified by Augustine's painful and pleasurable listening) that can split music in this way.

The relationship between a song's words and a sound's composed melody, or the sound of the voice and our experience and understanding of it, boils down to questions about what the experience of music is. These questions, as posed by Holsinger and as considered in *Body Music*, include: "What is it to 'experience' music? Where and how is music located vis-à-vis the persons who listen and react to it? How do we approach music as a sensual, passionate, and emotional medium, and how might we account for its widely varied effects on and interactions with human bodies?"[85] In contrast to a focus on words, speech, and writing, I consider my work with *Body Music* through Holsinger's evocative questions. I suggest that if we reframe musicking's core, understanding it as a constellation of corporeal activities and sensualities, we accomplish nothing less than a reconfiguration of the body's position in relation to sense and meaning making.

Emphasizing signifying sounds, as a semiological context begs us to do, skews the reality of the full event that is music and voice. In this context, sounds are selected, isolated, notated, and repeated. And in this dynamic, sound appears as the primary point, with the body and its actions—which create the sound—considered to be mere afterthoughts. That is, the body and its actions are considered as what Derrida—taking the term from Jean-Jacques Rousseau—calls supplements. Rousseau saw a supplement as "an inessential extra added to something complete in itself."[86] A supplement, then, is that which is secondary because it serves as an aid to something original or natural. Derrida offers writing as a prototypical example of this relationship: "if supplementarity is a necessarily indefinite process, writing is the supplement *par excellence* since it proposes itself as the supplement of the supplement, sign of a sign, taking the place of a speech already significant."[87] What characterizes the supplement, then, is its double function as both "substitution and accretion."[88]

Among the consequences of such misjudgments regarding "that which is complete in itself" and the supplement is our inability to fully account for the power of voice, sound, and music. In short, because we have been preoccupied with the codified sound, when a full event does not align with signifying sound schema, we do not recognize it as, say, music, and do not account for the effect it nonetheless may have on us. We listen for sound and are oblivious to the action. This omission leads me to wonder whether Rousseau's idea of originary lack—the notion that, by definition, a supplement is incomplete—is a result of semiological logic that excises selected sounds from the thick event and codifies them, with a profound result. However, what if vocal sounds were no longer made in the service of signs? What if words, music, and writing are supplements to something, but not fundamentally to signs? Add to this the idea that voice reflects experience, for example in its "grain," to use Roland Barthes's term.[89] Do these questions point to the impossibility of complete calibration within a semiotic system, an idea that the notion of supermusicality captures?

Looking to Jakobson and, to some extent, to Saussure and Derrida in an effort to understand and untangle the tension brought by notation into the process of composing *Body Music*, I recognized that it was not notation per se that had pulled the piece away from my prenotation intentions. The distinction, then, is not about the mode or format in which the most refined level of difference can be detected, bringing us closest to the original. In considering Derrida and Saussure, I realized that they both work on the same axis, dividing original from copy and the true from the derivative. Jakobson and scholarship on early negotiations of the relationship among words, music, sense, and notation led me to consider the value systems and principles that beg for relative fidelity to an a priori signified. As noted at the outset of this discussion about notation, these principles are not devised from the presumed virtues of the formats of speech or writing. Instead, they are derived from the values of the overarching paradigm, a paradigm that produces vocalization and listening that aspire toward fidelity.

In summary, we see that there are vastly different answers to the question of the relationships among action, sound, and notation—answers that point toward the challenges of establishing a hierarchy of those relationships. In a situation in which the mouth opens and closes, an utterance can be noted as /pa-pa/, the utterance is taken to signify paternal caretaker, and it is inscribed as "papa," many would single out the mouth's opening and closing as the surplus—that which was supplemental to "that which was complete in itself." From this perspective, the process of sounds forming words is understood as "that which is complete in itself." However, I suggest that such a view expresses a misunderstanding of what fundamentally takes place in the exchange between father and son, or in any other vocal exchange. I suggest, as in the meeting with Pollock's work, that "that which is complete in itself" in the exchange between father and son is not the expression of the action (which is understood through the filter of a form of codification), but the recognition of the baby's action.

To hint at what I will develop in the following chapters, I believe this tension comes from asking misguided questions about the material at hand. These types of questions arise from a logic that is driven by what I have called the figure of sound, which assumes that there is a standard, an a priori, against which to measure a given sound's or inscription's fidelity. Recalling how Rebecca Lippman's question (in the introduction) about the beginning and ending of a sound falls short within a framework of vibration and propagation, I suggest that a strategy establishing a hierarchical relationship between sound and notation—focusing on which captures difference more accurately—falls short within the investigation of music, voices, and human-made sounds. This frees us to ask: What if words and their sounds are supplements of something else—but not of an experience that awaits naming? Furthermore, what if we imagine sound, as *Noisy Clothes* suggests, as merely subsequent to and supplemental to action?

From Identifying A Priori Sound to the Process of Listening

In attempting to move toward a response to these questions, my own strategy lies in interrogating the distance between the experience, choreography, and anatomical action of voice, whether in purposeful linguistic or nonlinguistic utterance. In the same way that writing is a physical imprint and an impression of the writer, vocalization leaves a physical imprint and impression not only on the listener but on the vocalizer herself or himself.[90] That is, vocalization is both activity and experience, and any meaning we might derive from it is not separable from the experience. The specific word or sound communication *mama* might arise from a child's delight over seeing her caretaker, but it is equally likely to result from twice parting the lips while exhaling. Though on the surface the mimicosomatic reaction of lips and breath might be seen as reaction rather than communication, both iterations are equally communicative about that moment in time.

I propose that it is the action of voicing and the experience of voice that give rise to an inner corporeal landscape, which forms the basis of experience and meaning making.[91] In this case, it is the lips parting and the exhalation that express the child's situation at that moment in time. When we observe that speech or vocalization is never detached from bodily experience, we can no longer maintain that writing is a supplement of speech, which itself is imagined as a supplement of experience. Thus the meaning that arises for the above utterance, for example, is not an a priori meaning derived from the experience that is rendered through speech and/or writing—it is a report of the experience itself. The vocalization triggers an experience, which in turn creates the ground for experience and the meaning we derive from it.

Because of the misleading focus on the voice's sound, attempts at understanding and mapping meaning production related to voice have led to much misunderstanding. It is not only that the meaning to which this focus seems to point is a supplement, but, I will suggest, even the sound itself is a supplement. Shifting the focus back to the experience as producer of sound and, therefore, unmediated meaning helps us avoid these misunderstandings. Specifically, sound has been so powerful and seductive that it has directed our attention away from that to which it is indeed only supplemental: the action of the body.

Multisensory physical activity—including vocal sounds and speech—is experience. Affect and meaning are derived from that full experience. However, as we have seen in *Boomerang* and *Body Music*, we seldom allow the physical activity that sometimes is manifested through vocal sounds to exist without a constrained relationship to systems of signification. That is, if the vocal sounds do not conform to these systems, they are defined by a negative relationship to a given system's valued parameters: they are considered babble or out of tune. By placing sound at the core of a multifaceted event that includes physical activity, corporeal changes, and the sounds that arise therefrom, we unfairly impose inappropriate evaluative criteria on those sounds. For instance, the sound of the voice, and the way in which the instinct to notate renders listening selective, has seduced us into believing that the power of voice lies in its sound, while in actuality the sound is merely the tip of an iceberg, the entirety of which shifts us into affective states.[92] Singing happens before the sound; it is the action that produces the sound. Listening, then, takes place in the shared activity of singing—the shared actions of moving and being moved.

This perceptual shift away from considering voice as defined by its sound (and its relation to meaning) to voice's action and the action's physical impression on the audience's body calls to mind the perceptual shift invoked by Pollock.[93] His focus was on the motion rather than the motive, when he said: "A method of painting has a natural growth out of a need. I want to express my feelings rather than illustrate them."[94] Pollock assumes that manifestations of his being in the world, including the dripping of the paint from the brush held by his hand, are extensions of him and thus his direct, rather than indirect, expressions routed via signifier and its signified. What takes place as a result of the physical configuration of his body—the paint dripping from the brush held by his hand, which extends from the trunk of his body—is reconsidered apart from its prior designation as accident, spillage, or waste; it becomes gesture and therefore communicative in the same way that the oral gesture we hear as "mama" is communicative. Pollock pulls into the center that which is normally seen as accidental, because it is without a priori meaning.

Reading these scenes via Jakobson's work on "mama" and "papa" and scholarship on medieval performance practice's relation to notation, I have attempted to tease out the notion that vocal sounds, like dripping paint, arise out of the physical conditions we find ourselves navigating. Turning conventional wisdom as well as theories of sound and semiology on their respective heads, the mark that the paint happens to cause, and the sound that the complex physical activity that the voice happens to manifest through its vibrating vocal folds, are not the self-sufficient system of the natural presence but the surplus, or "an inessential extra added to something complete in itself."[95]

4

ALL VOICE, ALL EARS
From the Figure of Sound to the Practice of Music

Beyond the realm of music, what can happen when we move from operating under preassigned names and meanings to focusing on the new experiences offered by singular articulations? Drawing on an example familiar to everyone, for Romeo, such a shift in focus makes possible his experience with Juliet, which would have been unavailable had he approached her under her preassigned Capulet index. However, we also witness the reappearance of the process of naming: ultimately, Romeo didn't relinquish naming as a process of mediated interactions. Having experienced love with Juliet, and believing that she is dead, Romeo kills himself because he would rather die than live without love. The crux of the matter is this: Romeo fixedly named his experience with Juliet "true love."

Because Romeo, unlike his parents, was able to momentarily put labels and names aside and interact with a Capulet, he could have a meaningful encounter with Juliet. However, in the same way that the Capulets and Montagues saw each other only as adversarial forces, in the end Romeo believes that only Juliet can embody true love for him. Just as, at one point in time, the two families based their fixed interpretation of their relationship on a particular encounter, Romeo bases a fixed interpretation of the effects of love on a singular articulation that involves Juliet. That is, rather than arising from a dynamic relationship of coarticulation, the Montagues, Capulets, and Romeo believe that enmity or love are connected to particular signifieds and essentially inherent in them. Similarly, when dynamic musical or vocal coarticulations are fixed to A#, for example—that is, to a classical vocal timbre or a notion of the sublime musical work—we limit both the people involved and the possible outcomes of any future coarticulation.[1]

Thus far I have argued that music does not exist independently of, or prior to, specific articulations. Considering pedagogies of singing, the present chapter pursues a particular strand of this larger thesis. Specifically, it moves beyond interrogating dominant musical parameters to demonstrate how the impacts of such naturalizations of sound, music, and voice do not exclusively inhabit the realm of definitions but also have material and human consequences. Pursuing this investigation further, I take up chapter 3's discussion of clinging to music's assumed building blocks of sound, and of how the notion that there is something stable about the concept of sound creates a false understanding of music's ontology. Ultimately, I argue that it is only through conducting our interactions with music outside the framework defined by these preassigned labels that we may begin not only to appreciate, but also to learn from, the multifaceted and sometimes contradictory experiences we have with music.

It is now a truism that the value and status of a given musical work are not self-contained and autonomous but arise within social, cultural, historical, and political spaces. However, what requires further investigation is how performers who must uphold the values and status of the musical work concept are subsequently affected. In pursuing this question, I observe and experiment within the practice of Western classical vocality. Moreover, I examine how given naturalized notions of music, such as sound and the related work concept, manifest in musicians' training. Vocal pedagogy, as the shaping of a human body optimized to deliver the sound concept, thus corroborates naturalized conceptions of sound, music, and voice. Therefore, while on the one hand this chapter attempts to disentangle the sound concept from the work concept, on the other hand it aims to make explicit the intricacies of the human labor required to hold the work concept together. Therefore, this chapter considers how given ontologies of music and voice structure our understanding and practice of listening.

Voice in the Service of the Work Concept

The Romantic musical work, or the "work concept," described by Lydia Goehr as signifying autonomous music as a privileged object of study, has accumulated so much traction that it continues to dominate the analysis of both present and past music.[2] Leo Treitler points out how this modus operandi causes anachronistic analyses, interpretations, and methodological applications.[3] Gary Tomlinson offers a crucial bottom-up perspective on rethinking historical work by grounding theoretical and analytical frameworks in the examination of the social, cultural, and spiritual spheres in which the music and practice arose, as opposed to gathering and synthesizing a larger number of data points from which to derive analytical tools.[4] These are examples of the intriguing work that arises from the complexities with which scholars are faced when taking into consideration the true dynamism of music.

Despite the seminal and intriguing work that has arisen as critique of the work concept, I suggest that it is important to expand that critical lens to focus on what a knowable and recognizable piece of music is. More specifically, how is vocal labor recruited to maintain a piece's identity and status?

In thinking about how contemporary performers, as thinkers about music, are shaped by the work concept, it is productive to consider timbre in relation to musical genres and the works that exemplify these genres. Among other aspects, the intactness of, for example, a given opera—the ontological sameness between the notion of an operatic work and its actual performance, and the perceived identification of an operatic performance—depends on, for instance, a classical opera performer's ability to deliver sounds that match tacitly agreed-on timbral ideals. Thus, minutely detailed timbral adherence, an element that is often indicated in very general terms (for example, *sotto di voce*, meaning "with a soft voice"), drives the perception of the work's status and directs performers' training and delivery.

Musical genres are generally recognized within a few seconds, based on timbre.[5] If the formal parameters of a genre are fulfilled but the timbral aspects are not, the status and intactness of the work in a particular instantiation—that is, the extent to which the work remains itself—are called into question. That is, when Luciano Pavarotti sings "My Way," its genre designation as a torch song might not be as clear as when Frank Sinatra sings it. Or when U2's Bono sings "Ave Maria," it might not be recognized by classical music aficionados as worthy of the same contemplation as a classical singer's rendition. That is, the work's retention or loss of a particular status rests not only on the composition and on listeners' ability to recognize it as that composition at a root level (in this case, as "Ave Maria"). Nor does it rest on audiences' agreement as to the status of the composition (that "Ave Maria" is a religious piece in the classical genre). A piece's retention or loss of a particular status also depends on the timbres used in the work's performances.

Consequently, the performer's labor goes beyond the careful production of beautiful timbres and individual works. She is required both to be an ideal listener who knows and listens within the frame of the work concept—that is, someone who recognizes the particular subtleties, often established and maintained through unspoken conventions that hold the concept together—and to develop the ability to deliver these musical subtleties to an audience. In other words, to maintain the work's autonomy requires dedicated human timbral labor. And the constant need for this labor is one area in which we can see that music's status, meaning, and definition are not a priori.

In the context of her reflection on the Romantic work concept, Goehr discusses extensively how "the work-concept began to regulate a practice at a particular point in time."[6] In the same way that sound is one of the parameters of music that I argue are naturalized, the labor that goes into offering and maintaining music's fine-grained nuances is crucial to keeping the work concept intact. As Bono's performance of "Ave Maria" exemplifies, the piece's identity breaks down if it is not performed with the expected timbre. I have come to think of this minute labor as timbral alignment with the given genre. The work practice, then, regulates musical practice, and hence the bodily practices of both musicians and listeners.

The structuring of the individual sounds and timbres that are required for a work to retain its identity also takes place within the dynamic of the power structure that is the work concept. In the same way, the notated page and our expectations regarding all musicking resources are depended on to realize the intactness of a particular work, and it is assumed that features such as the timbre of, say, a classical vocal production function "as neutral conduits, as instrumental rather than substantive parts of the social relationship."[7] In fact, as I will discuss in detail below, because the Romantic work concept rests on its assumed a priori status, the concept wavers and falls apart if the performers' hard-earned skill in presenting timbral alignment is revealed.

The appearance of timbral alignment as an instrumental conduit rather than as a requirement for the intactness of the work feeds into and is enabled by pedagogy that reproduces these timbres and their attendant ideologies and, as Frances Dyson puts it, is "ontologically separate from" the need to support the idea that a work's identity as itself "exists prior to and outside of its affiliation with the technology."[8] But the "politics of voice," writes Amanda Weidman, and the "ideology of voice" perpetuated by the work concept "are also a mode of discipline—embodied and performed—through which subjects are produced."[9] In other words, the work concept is technologically enabled. It is, in part, the limited awareness and examination of this dynamic that has led to the curious pedagogy we see in classical vocal practice—a pedagogy that oscillates between metaphorical language and scientific explanation with aesthetic and sound-based impetuses.[10]

Concepts' Impact on Vocal Pedagogical Frameworks

Aspects such as timbre receive attention to maintain the work status of a given composition, set the agenda for performers' training, and establish the relationships between a given ontology of music and a given power dynamic within pedagogy. The work status of an entire tradition (for example, classical opera) complicates the need to acquire the ability to perform it and the question of what exactly it means to undergo the process of learning it. And performers of classical music are largely trained in environments that do not recognize the constructed nature of the work.[11] One might go so far as to hint at the dependence of the very livelihood of these institutions on the unquestioned stability of the work concept, enabled and maintained in large part by its attendant performance practice. How is voice conceptualized and taught under these tremendous pressures, which produce so many inconsistencies in spoken and unspoken norms and expectations?

While voice has historically been recognized as an area of investigation for medicine and science, classical vocal pedagogy has only relatively recently showed a marked interest in self-conscious scientific framing in the service of aesthetic goals.[12] Most notably, in the mid-nineteenth century Manuel Garcia Jr., the inventor of the laryngoscope, was the first singer and voice teacher who systematically studied the voice from an exclusively anatomical point of view. With his successful students and two widely disseminated volumes on the science of the singing voice, *A Complete Treatise on the Art of Singing*, classical vocal pedagogy made a tremendous perspectival shift, as it began to include considerations and explanations of the voice from anatomical and functional points of view.[13] However, despite Garcia's influential contribution, vocal training today remains grounded in science but replete with nonscientific imagery. We may find a strong reaction to nonscientism in William Vennard's now-classic vocal manual, *Singing, the Mechanism and the Technic*. A book that gained an authoritative position in the vocal pedagogical community, Vennard's 1967 text advocates science-based singing and teaching:

> There are certain fundamentals of the science of acoustics that all voice teachers should know. Too many have a superficial acquaintance with the vocabulary and dress up their pedagogy with terms that sound im-

pressive but lead to confusion when the student, in the course of learning physics, finds out what the words really mean. Scientists laugh at the imagery with which the voice teacher tries to express himself, and they have even more justification when the imaginings are clothed in words like "resonance," "fundamental and overtone," "sounding-board," etc., which have specific denotations in the laboratory. Some teachers take the position that, since much of singing is still unexplained scientifically, and since it lies below the level of direct conscious control anyhow, it is better for us to avoid these discussions, to admit that we are unscientific, and let it go at that. . . . It is true that singing can be taught entirely by abstract, more or less emotional appeals to the entire personality of the student, but I cannot escape the conviction that many times more direct methods bring quicker and better results.[14]

Vennard expresses disapproval of teachers who do not draw on or correctly employ scientific or medical knowledge and language. Furthermore, he posits that while it is possible to achieve satisfactory results from a metaphorical instructional basis, there is more complete and concrete knowledge about the voice and its functions available in the medical and scientific community, and it seems counterintuitive not to make use of it. But ironically, after chiding pedagogies that default to nonscientific language and imagery, and immediately after his highly technical introductory chapter on acoustics, Vennard fails to follow up with a lesson that directly applies this theoretical knowledge. Instead, he advises the singer: "It helps to imagine that you are a marionette, hanging from strings, one attached to the top of your head and one attached to the top of your breast bone. This keeps the head erect and lifts the chest, allowing the pelvis to just 'hang' in position."[15] This passage and others in Vennard's text illustrate classical vocal pedagogues' struggle to connect physiological knowledge with singing pedagogy. Indeed, theoretical fluency in the medical sciences does not easily translate to elegant pedagogy.[16]

Therefore the use of metaphorical approaches persists. Such approaches are metaphorical in the sense that they are indirect—for example, they emphasize enjoying nature or acting natural rather than the physical techniques involved in singing—and their choice of words is often ornamental. Take, for example, the instruction to "inhale as though you are smelling a rose," a notorious axiom in vocal circles. What action does this instruction aim to produce? Does it encourage breathing through the nose rather than the mouth? Might it be promoting the introspective stance assumed when noticing the fragrance of a rose, giving rise to a relaxed rather than tense general posture? Is imagining a pleasant fragrance meant to produce a slight smile? Or might it simply be that this pleasant, ladylike, and generic metaphor of the singer's outward presentation was the original goal of the exercise? What specific insight this instruction actually offers the singer is unclear. There are a number of similarly vague instructions in common use, such as: "Breathe from your stomach," "Sing like you're pulling a thread through your nose," "Focus your tone," "Breathe from the ground," "String your notes together like pearls on a necklace," and "Imagine you have an airplane in your larynx."

These metaphors reflect the tension that takes root in vocal instruction as it is pulled between the aims of aesthetics and those of the medical sciences. In part because voice serves in this context as an aesthetic tool and is ideally appreciated as an a priori aesthetic entity (as expressed by the saying "a born singer"), it is never permitted to cast itself as an exclusively anatomical function and technique. Additionally, while the corporeal nature of the voice has been universally recognized, complexities arise in translating anatomical knowledge into voice lessons. In part, these challenges are related to segments of the vocal apparatus made up of involuntary muscles.[17] (The other challenge, which I will discuss below, relates to musical notation.) I suspect that pedagogical practice's continued leaning toward the metaphorical realm is only partially explained by voice's entanglement in an aesthetic designation combined with an anatomical reality. In this chapter, I suggest that what trips up classical vocal pedagogy is the fact that while we possess tremendous knowledge about physiology, we still believe that music and voice consist of sound, not physiological matters. It is this belief that prevents vocal pedagogy from creatively working from a complete base of knowledge about voice. Two ramifications follow. The first is that singers take instructions from the notation that arises from a premise of sound. The second is that vocal pedagogical practice is truly built around voice as sound.

Traditional Western notation is based on abstract signs, which carry a host of implicit and explicit information that a performer is supposed to translate into action. Within music notation there is already a premise of abstraction. That is, a sign is presented, but what that sign means for the singer is not inherent in the sign; rather, it must be learned. The challenge of the abstraction that makes up a score is that it is a text-based abstraction of the sound. Since singing is action, the score thus presents an enigma instead of offering straightforward instruction. The singer must project herself into the sound, somehow working herself back in time to solve the puzzle of what the body should have done to produce that sound.

Despite intermittent references to the anatomy of the vocal apparatus, the conceptualization of singing through sound fails to present a path by which the singer is intuitively prompted to access the mechanisms and actions of singing. Furthermore, the sung sound of the word *mio* on the note A is treated as synonymous with the A represented in a score, or the knowledge that this A equals 440 hertz. In other words, the vocalizing signifier must produce a sonic signified that is itself bound to a stable definition. As a result, traditional vocal pedagogy rests on the assumed correspondences between the abstraction of sound (the musical note), the sign (the note designation A), the sound wave's oscillation, the existence of this ideal sound in the voice, and the assumed ability to easily call on one of these correspondences.[18] However, as I have suggested, to sing is to carry out the sets of activities that will produce or give rise to a pitch and timbre required within a certain genre, akin to musicking.

From the perspective of voice as action—viewing a particular sound as merely the outcome of a given action and its material energetic transmission—imagine an ice skater's instructions as notated abstractions of the sounds produced by her movements. I suggest that the relationship between a singer's job and the notated score is fairly similar: in reality, the singer's task is to carry out a set of actions, which I have called an inner choreography, analogous to the skater's movements. This inner choreography gives rise to the sound we call voice. And it is with the abstraction of this resultant sound that traditional scores are concerned, even though the singer's work lies in the realm of action.

Furthermore, because the premise of the stable work concept—from which an equally fixed notion of voice as sound derives—is the concept's inner coherence and its presumption of a priori workness, the carefully assembled timbre contradicts this picture. In this regard the singer's unrecognized labor is twofold. First, she or he must determine and execute the general actions required to produce the sound indicated by the score. Second, she or he must determine and execute the precise timbre required to express the genre and keep the work's identity intact—as, for example, an opera aria. It is crucial that this labor takes place without drawing attention to itself. As a result, the human labor that goes into maintaining timbral exactness and keeping the work concept intact is another naturalized area in music.

In her discussion of music technology, quoting Jonathan Sterne, Dyson identifies this uneasy relationship between labor and the effect or outcome it enables: "Acousmatic or schizophonic definitions of sound reproduction ... assume that sound-reproduction technologies can function as neutral conduits, as instrumental rather than substantive parts of social relationships, and that sound-reproduction technologies are ontologically separate from a 'source' that exists prior to and outside of its affiliation with the technology."[19] Dyson's critique of the assumption that "sound-reproduction technologies can function as neutral conduits" is helpful in thinking through the example of the singer's body, training, and labor in relation to performance of the work. While varying ideas of the work necessitate differing kinds of labor by performers, the Romantic work concept calls for, and arguably stands or falls depending on, the concealment of the intense labor involved in executing and accomplishing the work. The logic upheld by not acknowledging this labor is as follows: if minute changes in timbre can thrust the composition into a different conceptual category, then what is a priori about the work?

In my own training, for example, working under these conditions produced a feeling of intense contradiction. That is, knowing the sort of vocal timbre I was asked to produce, but not how to produce it, left me feeling that I had been sent on a wild goose chase after this sound without the means to reach it.[20] More precisely, I felt as though the sound existed outside of me, an icon hovering just in front of my face. While my knowledge of the desired sound seemed crystal clear, the set of steps necessary to produce it was less so. In fact, I felt that the teaching I was given somehow removed me from the process. Viewing the rose-smelling exercise through the framework of sound-based singing, we see that the exercise derives from what the singing person looks like when the sound is heard, which is indeed after the fact, rather than from the motions that constitute the actions that produce the sound. I propose that because people focus on the sonic product they see and hear—that which is the consequence of inner choreography's prior action—they index the event with records that describe the after-the-fact state, rather than those that produce the desired sound.[21] The point I attempt to address in this book is that we should note this condition and be aware of its effects, of the broader ramifications of this type of vocal production, and consequently of the vocal bodies it creates.[22]

A Sound-Based Voice Lesson

Much of my thinking, and much of the reconsideration of voice presented in this book, comes from an experimental, performative practice designed to test the two hypotheses of voice as sound and voice as action. The hypothesis of voice as action came to me in the context of teaching voice and through reflecting on my own practice. I wondered if thinking about voice as action could lead to a different relationship among a singer, her body, and her sense of her own voice. Through a series of pedagogical and musical experiments, I tested this hypothesis in an attempt to discover whether the challenge of connecting vocal physiological knowledge with singing pedagogy, and the sense of disconnect described by many singers, could be mitigated by shifting the emphasis from sonic outcomes to physical actions.

A sound-based notion of singing offers a twofold pedagogical model. First, pedagogy centers on acquiring and internalizing knowledge of the sonic ideal that feeds into the work concept: it teaches us to recognize this ideal in other singers' voices, to recall it internally, and thus to recognize it when it is present in our own voices. That is, it embeds the sonic ideal in the singer's internal ear—which, as we recall from chapter 3, may be understood through the notion of the "automatic functioning of power."[23] Second, in the step that follows, the student attempts to change her voice to match this ideal, referring to the ideal as though it were a key. In other words, the goal is to cause the body to produce this preconceived sound concept via an idea of the sound, rather than through facilitating the bodily stance and actions that could yield that sound or another one. Keeping the sound ideal in the center gives priority to matching that sound. Furthermore, and necessarily, sound production takes place through simultaneously singing and evaluating the result, forcing the singer to engage in production and evaluation of an indexical sound rather than in singular, unrepeatable particular sounds. Chapter 3's discussion of Richard Serra's *Boomerang* describes this dilemma and its schizophrenic call to action.

The process of sound-based singing that supports the work concept is most commonly taught by dividing each lesson into two parts.[24] Clear sonic goals are laid out for the singer as the lesson proceeds, starting with a warm-up and moving into repertoire work. During the warm-up, the teacher commonly presents a pitch set of five descending notes on the fifth to first scale degrees (for example, the notes G, F, E, D, and C) and asks the student to sing these pitches on a selection of vowels or phonemes such as "ah ah ah ah ah," "ma ma ma ma ma," or "mi me ma mo mu." This last, ubiquitous combination features the five basic Italian vowels (i, e, a, o, u) paired with a clean, easily resonating consonant, *m*.

The utilitarian goal of these exercises is to match pitch, vowel, and quality or timbre. The aesthetic goals of this type of training include broadening the vocal range, learning to correctly shape vowels and consonants (according to the standard conventions in a given repertoire), and facilitating ease and uninterrupted flow in the vocal line. After the warm-up, most commonly dedicated to checking vocal alignment (with the aim of a unified timbre along the entire register) and working on problem areas, the typical voice lesson moves into repertoire work. Here we see the same sonic expectations that are featured in the warm-up. Work with repertoire continues along the same lines as the vocal exercises, while presenting additional challenges in terms of pitch range and intervals, musical demands (rhythm, accents, and so on), combinations of vowels and consonants (and the particular musical parameters within which they are combined), and other interpretive demands.

Sound-based voice pedagogy focuses on sonic accomplishments. The various parameters and their relative importance are set in a clear hierarchy. The number of parameters to which the student is able to attend also marks her or his accomplishment.[25] Additionally, the parameters are clearly ordered according to their importance, from the correct pitch and duration of notes and syllables to subtleties regarding the delivery of consonants, movement from one consonant to the next, and so on. Overall, this type of lesson is based on a fundamental belief that the pedagogical goal is to move the voice as close as possible to the sonic ideal set by the genre. In short, what I have come to see as sound-based singing revolves around conformity in regard to sound. The bodily activity that accomplishes this conformity is considered secondary—or not considered at all.[26]

As the teacher corrects, often by demonstration or verbal description and sometimes—but less often—by corrective physical instruction (such as "straighten your neck"), the student's ear is trained to hear the same things that the teacher's hears and to catch any anomalies that fail to line up with the sonic ideal. Even if the correction is physical in nature, it is implemented with the desired sound in mind. These exercises are meant to gradually move the student's voice closer to the sonic ideal through imitation. Such basic training takes, on average, around six to ten years, and professional singers continue to engage with voice teachers and coaches throughout their careers to ensure that they maintain this sound.

Although the activities involved in singing seem to take place simultaneously, we can deconstruct this multifaceted act into the following sequential events. Work-based singing begins with holding the idea of sound in one's mind—an ideal predetermined by generic or stylistic conventions. The desire to match this sonic ideal is an impetus to an action, which produces a result that is evaluated according to how well it aligns with the preconceived sonic notion. This means that while she sings, the singer constantly listens to and evaluates her sound in relation to an external sonic idea, akin to the *Boomerang* performer discussed in chapter 3. Because the catalyst for the act of singing is the motivation to match preconceived sounds, it results in an act of re-creation rather than creation. And because this endeavor involves matching an a priori sound ideal, the result—always a copy—will inevitably fall short.[27] Moreover, because the act of singing primarily aims at the produced sound's relationship to an ideal sound, it results in a disembodied action, with the singer spending most of his energy on evaluating his sound relative to the ideal, rather than focusing primarily on sonic production or on the resulting sound's qualities and virtues.

Admittedly, learning to sing according to the classical Western opera model is a rarefied practice, and an experience shared by a relatively small percentage of singers. However, this rarefied tradition exemplifies and rehearses sentiments and values that are not limited to its milieu. And while classical singing is hardly a universal practice, it has disproportionately affected academic discourse on singing. Therefore, while classical vocal pedagogy provides an extreme example, it also offers a model for thinking through this issue: the belief in and practice of the sound-based notion of singing are not limited to classical singers and pedagogues. For example, even when you sing along with a recording, nobody is telling you what to do, but a sound-based perspective nonetheless orients your activities. Anytime we operate under the ontological impression that sound and music are stable and knowable, it is a sound — the way in which we have come to understand and name a singular, unique articulation — that guides our singing.

From the study of voice-as-sound lessons we can extrapolate a general model: whether our relationship to a given sound takes the form of conformity (imitation) or repulsion (creating a nonimitative sound that is nonetheless derived from a relationship to the initial sound), it is the sound as we understand it, filtered through its name and related meaning, that we seek to reproduce (rather than moving into new and singular articulations).[28] The ideal voice can be of many different kinds — perhaps even one that refuses to cast itself in sonic terms, but that nonetheless contributes to the dynamic of sonic and vocal ideals.

For example, Annette Schlichter observes that the Kristin Linklater method for actors' voice training "offers the performative production of a form of natural vocality that allegedly expresses an 'authentic self.'"[29] Its vocal ideal is what Linklater terms the "natural voice," which she promises to "liberate" through vocal training. This voice "serves the freedom of human expression," freedom from the corrupting influences of culture and therefore is understood as a truthful expression of an authentic and universally understood human self.[30] Although Linklater's language is radically different from that of the opera world, it is interesting to observe that her ideology is equally strong. While she self-consciously refuses to name specific sonic ideals, her methods are clothed in vocabulary such as "freedom," "authenticity," the "natural voice," and a desire to "educate the voice into the union of self and body."[31] Furthermore, these ideals are believed to be achieved primarily through intimate engagement with

Shakespearean language in combination with long-term, on-site, committed work with Linklater or an instructor she has certified. Hence, while vocabulary that refers to specific sounds is avoided in this vocal training and practice, the architecture that forms the practice contains elements similar to those of classical vocal pedagogy: a specific repertoire and exercises believed to bring forth the natural voice, vocal technical specifications that call for resonating fully in a room without the aid of amplification technology, and finally an intimate working relationship with a person authorized to teach the practice.

While the Linklater method avoids the explicit naming of sound qualities, it is a teacher's call to determine what is the student's natural sound. So, while the sound-determining process appears more forgiving than it does in classical vocal pedagogy, the structure wherein an authority names the sound is actually the same. Linklater and her instructors, who self-consciously avoid naming sounds (since that would undermine the concept of each person's natural and free voice), still act according to an inner sound barometer that allows them to recognize the free voice. That is, this vocal method is rooted in voice as sound.

Structures of power are funneled into sound ideals. This intensification takes place at the base of vastly different vocal practices and training scenarios where aesthetic ideals, training structures, and groomed practitioners intersect. Given that we hear and practice within given vocal structures, how can we produce a critical distance from these contextually naturalized sounds and practices? Schlichter's term "voice work" is productive here.[32] In short, it names the constructedness of both sound and practice, helping to create the critical distance that allows us to examine the labor that produces and maintains these sound ideals. Understanding that voice work is indeed an "embodied discourse-practice," to draw on Schlichter again, "calls for a denaturalizing critique of an ideology of vocality."[33] This ideology can, in the case of classical vocal production, uphold and support the appearance of the immanent work; in Linklater's institutionalized actors' vocal training, the belief in a natural voice, freed from culture, is imparted.[34] The term *voice work* also helps us dig deeper into ideologies such as Linklater's that purport to free the voice from technology, even though speaking to and with a room such as the specially acoustically designed Shakespearean Globe Theatre is, of course, as technological as the use of amplification.[35]

Key to understanding the power performed through voice is an understanding of the heavy vocal labor shaping every utterance and informing every instance of listening to voices. Because the natural voice has never existed, the qualities we essentialize from the voice to the vocalizer (including authenticity, subjectivity, truthfulness) are also not natural, but learned and performed. This leads us to ask whether, when voices are overtly trained within distinct and self-conscious vocal traditions and every quotidian vocal interaction is politically, ideologically, and socially generated, "one's voice belong[s] to oneself at all, and if so in what sense?"[36] Furthermore, as Mladen Dolar notes, "voice is not a primary given which would then be squeezed into the mold of the signifier, it is the product of the signifier itself, its own other, its own echo, the resonance of its intervention. If voice implies reflexivity, insofar as its resonance returns from the Other, then it is a reflexivity without a self—not a bad name for the subject. For it is not the same subject which sends his or her message and gets the voice bounced back—rather, the subject is what emerges in this loop, the result of this course."[37]

Dolar warns that fetishizing the voice creates a barrier between us and our ability to critically analyze the voice as "the product of the signifier itself." He argues that "singing, by focusing on the voice, actually runs the risk of losing the very thing it tries to worship and revere: it turns it into a fetish object—we could say the highest rampart, the most formidable wall against the voice."[38] The voice as sound experimentation confirms Dolar's observation. However, *Noisy Clothes* and *Body Music* suggest that there is a way to interact with the singing voice without fetishizing it. I will now discuss this idea in more detail through experimentally testing the voice-as-action hypothesis. Thereafter, I will propose that we take Dolar's insight even further and ask whether we can relate his statement about voice as a product of the signifier to the discussion about my proposed reversal of supplement and core (sound being the supplement and action being the core) in chapter 3. In the dominant view, in which sound is the core, perhaps it is not the voice or logos but the sound that is the fetish object of the voice, inhibiting us from knowing the voice more closely and understanding more about its effects on singers and listeners.[39] Moreover, if we were not deafened by sound, would music appear to us not as an ideal and fetishized sound that helps maintain the work concept, but as a compound manifestation of performers' bodies?

All Voice: An Action-Based Voice Lesson

Applying a shift in focus from the finished product to the action that produces it (exemplified by *Noisy Clothes* and *Body Music* in chapter 3) to the dilemma of vocal epistemology and heuristics described above, we understand that a song or any piece of music is created through the collective musicking—the cooperative participating in sound—of each person and community. We can apply these insights to strategies of vocal instruction and, more broadly, to the ways we conceptualize singing. As I explained in chapter 3, singing is the action that takes place before the sound is materialized. In the words of Catherine Fitzmaurice, music "has no location in the body except when it is in action, sounding."[40] In addition, sound is not an a priori index. In relation to singing and its teaching, we may recall the analytical shift surrounding Jackson Pollock's work in which emphasis was transferred from the finished painting to the actions that created it. Similarly, vocal instruction's emphasis could be shifted from the resultant sound to the actions that produce it. And maintaining that sound and singing are singular, unrepeatable instances and articulations, and that we can engage with them only at that level, would preempt all efforts to reproduce a named indexical that has since vanished.

I have taught voice since 1994. Through my intense engagements with the teaching and learning process I not only began to question conventional methods, but also to experiment with alternative frameworks and methods. Needless to say, my theoretical work on voice has influenced my practice, and vice versa. If pedagogy were to cultivate the concept of sound that I articulate in this book—the notion that sounds are actions leading to singular, unrepeatable articulations—what would such pedagogy be like? First, instructions would be action- rather than sound-based. Second, this pedagogical model would create scenarios in which the student is prevented from judging the sonic outcome and from adjusting her voice according to a feeling of success or failure in relation to the outcome.

Developed through weekly interactions with students, and with the help of their feedback, my action-based voice lesson begins by giving the student a series of aerobic physical exercises that do not involve producing vocal sounds—for example, running around, jumping up and down, and so on. I then resize these activities and transpose them to the anaerobic realm, but I work to maintain their general level of energy. So, for example, the activity of moving the arms while extending them away from the body's trunk is slowed down and moved closer to the body. These exercises are intended to initiate the use of the vocal apparatus (the entire body), and also to guide the student's concerns away from the voice per se to fully physically engaged activity. When we finally start adding sonic components, it is crucial to prevent the student's attention from moving from the action to the sound. To this end we work toward three specific goals: (1) building on actions that can also yield vocal components and make the vocal sound appear to be either a form of play or an artifact of the general exercise; (2) keeping the student so busy with actions and movements that she does not have the time or energy to pay attention to the resultant sound; and (3) transferring a bodily stance from the instructor to the student.

An example that includes all three components is the exercise in which we imagine we are standing at the edge of a very high cliff with our toes hanging off the edge—just at the tipping point, beyond which there is only a drop into empty space. The student attempts to avoid falling by holding herself up and back with her entire musculature and energy (objective 2). This situation is simulated several times. First I stand very close to the student, so that the sides of our bodies are touching, and I often put an arm around the student. The goal is to invoke shared bodily energy by breathing together (objective 3). We enter this situation and position a few times without any vocal sounds. When I sense that the student has fully embodied this physicality, I invite her to imagine the excitement of being at such close proximity to the cliff's edge and to release the kind of vocal sound that would come from being in such a situation—a sort of scream, or an exalted sound (objective 1). During this process it is key that the student does not judge and correct her sound as she creates it.[41]

As we slowly move toward more conventional singing, the activities that have yielded general phrase contours as a result of actions are now repeated. Here the key is to avoid offering a pitch or pitch set that the student is asked to match. Also key is to maintain the centrality of activity, which only secondarily happens to yield a pitch. The pitch contour discovered through an activity is slowly transformed, or transposed, into a phrase of song.

To summarize, the principles of action-based pedagogy include: (1) avoiding listening for an ideal sound and judging the sound produced (in other words, sound is not a goal); (2) avoiding singing and correcting one's sound simultaneously; and (3) building a bodily choreography that yields sound and slowly scaling back the outer manifestations of that activity, such as large arm movements, while maintaining its intensity level. These guidelines invite each singer to discover his body's sonic possibilities by exploring its potential for action. Only subsequent to a particular action may he discover the pitch and timbre of his body as it undertakes that action.

Sound-based vocal pedagogy focuses on what the voice sounds like in the imaginary, rather than on how it is realized in and through the body. Instructions such as "inhale as though you are smelling a rose" call for the re-creation of the outer appearance of a still-undefined action, and not necessarily for the re-creation of the action that led to the sound. In action-based singing, in contrast, there is no preconceived sound that singers must attempt to match. In fact, learning to sing within this model aims largely at unlearning ingrained patterns, many of which involve listening to oneself while singing, to measure one's own sound against a sonic ideal. Instead, in action-based singing, one concentrates one's energies and attention on the actions involved in singing, which are indeed what give rise to sound. Then it is impossible to make mistakes: all sounds are, by definition, what they are supposed to be—since they are supposed to be whatever they turn out to be. Such a shift in the conceptualization of singing—from sonic to active—has enormous ramifications for our understanding of the exchange that occurs between listeners and singers.

If, at its core, singing is akin to a falling tree, classical singers train for decades to perfectly reproduce the abstracted idea of a sound, without ever realizing that that idea arose from a tree falling. In action-based pedagogy I try to facilitate the safe falling of the tree and to allow for the experience of the thick event, and the witnessing of the effects of that thick event. And, despite my dust-burned eyes, choked-up nose, and thumping body, I sense that singing and listening are not confined to the audile register but rather permeate us. At its base, the ontology of singing and listening is material practice.

All Ears: Multisensorial Listening

If singing is not sound but action, what takes place when we listen? If singing is no longer intended to re-create a sound that has been signified and indexed a priori, then listening is no longer a matter of recognizing the accurate production of the a priori signified. When we reconceive singing as action rather than sound, the definition of song is no longer stable; rather, it continuously emerges according to an unfolding process and a play involving multiple parts, including the material sound source: the materiality through which the sound is transduced and the materiality of the listener. If singing is reconceptualized as a total-body activity, listening—or the receptive engagement with singing—must follow suit, as will the ontologies of music, sound, voice, and listening.

In sound-based listening, above all we attempted to name the effects of each singular articulation. Thus we identified "G#," "G# minor scale," "aria," "opera," "lyrical tenor," "dramatic soprano," and "Western classical vocal sound" as strictly defined categories on which we must call in our attempts to replicate their effects. Consider this example: as the feeling of reverence arises in response to an aria, we ascribe worthiness to the aria. In other words, we assume that there is something inherent in the aria that gives rise to reverence, just as Romeo assumes that something inherent in the thirteen-year-old Juliet gives rise to once-in-a-lifetime love. Like Romeo, who believed that his newfound experience of love could be derived only from Juliet, when we name the effects of a singular and particular articulation we begin to seek the signifier, rather than holding ourselves open and ready for other singular and particular articulations that may become equally meaningful and important to us. Seeking to replicate that sense of reverence, for example, we seek out the particular aria that produced it. To listen beyond the familiar categories and their prescriptions, and thus to listen beyond the paradigm of voice as sound, we must think beyond the limiting categories to which listening is usually bound.

These categories are exemplified by Michel Chion's "three listening modes": "casual listening," which gathers information about the sound's source, from the owner of the voice on the telephone to the gravel underfoot; "semantic listening," which uses "a code or a language to interpret a message," from phonemes to spoken words; and "reduced listening," which concentrates on the sound itself, not as a "vehicle for something else."[42] While Chion considers listening to be more complicated than hearing, we could extend even his definition of listening to include an active contribution to the material circumstances surrounding that which is sensed. As a material entity, I partake in the material propagation that I understand as hearing or listening. If I considered my sonic experience from the perspective of its participant and cocreator, aspects of it that previously seemed liminal might in fact reveal themselves to be ordering and distinguishing properties of the experience.

There are some well-known examples from the history of music and sound technology that may prove helpful in taking further this thinking about listening. In these examples, compensation is sought for an eardrum's diminishing sound transmission, or hearing is explored in the process of developing sound technologies. For instance, it is said that as Beethoven's hearing slowly declined, he began to make fuller use of the multisensory dimension of sound. He supposedly held one end of a wooden rod between his teeth while the other end touched the piano's soundboard. In this way, the piano's vibrations were transferred from his teeth to his jawbone, to his skull bone, and finally to his inner ear. It is on this principle that current jaw-conducting hearing aids were developed.[43] In another purposeful transmission of sound, Beethoven removed the legs of the piano, transferring the soundboard's vibrations to the floor. Sitting on the floor, he would receive vibrations through his ear trumpet (the only hearing aid available at the time) and through large areas of his body such as his buttocks and thighs. At times he would even rest his head on the floor to facilitate transmission from the floorboards to his cranial bone. Beethoven also experimented with his ear trumpet. Instead of simply holding the trumpet open to the air, he would aim it at the sound source, or place the trumpet directly against the source. All of this reminds us that objects contribute to the material reconfiguration of sensing and the propagation of sound.

The senses, as Steven Connor and others tell us, never operate alone; we intuitively seek out situations in which we may take advantage of their commingling.[44] In the same way that Beethoven supposedly used bone conduction to have a more visceral contact with musical vibrations, Thomas Edison held wood between his teeth to enhance his hearing. Connor is interested in Edison's claim that this allowed him to sense overtones that he could not have otherwise experienced, given his compromised hearing: "The sound-waves then come almost direct to my brain. They pass through only my inner ear. I have a wonderfully sensitive inner ear . . . [that] has been protected from the millions of noises that dim the hearing of ears that hear everything. . . . No one who has a normal ear can hear as well as I can."[45] Here we see Edison's belief that shielding himself from the sounds that indiscriminately enter the ears of normatively hearing people allowed him to configure his material body to transduce sounds in a heightened way. Edison even privileged this mode of hearing, implying that because he was not distracted by the constant sound that fills the world, his sensing and hearing were paradoxically liberated.

Others have also developed theories about and practices related to the idea of a bodily listening. The French anthropologist Marcel Jousse observed that the "reception of [an] event involves the sensorium in its entirety, entwining proprioception with various forms of synesthetic experience—the patterned interconnections of touch, vision, hearing, smell."[46] Drawing on Jousse's thought, Charles Hirschkind discovered that even some people with intact hearing deliberately engage with sound using more than just their ears. Through ethnographic work with people who listened to religious sermons on cassettes, Hirschkind found that their listening activity was not limited to information comprehension, as the sermons' power lay beyond the factoids they delivered. Privileging multisensory communicative modes is particularly central to Islamic listening practices, even those that are mediated through technology. In his study of a group of Egyptians listening to Islamic sermons on cassettes in 1990s Cairo, Hirschkind found that what he describes as an "ethical listening" practice necessitated a particular stance: a deliberate positioning of the entire body.[47] That is, listening instructions are not about what sounds to listen for, but about preparing the body to move into a psychic and physical position that will allow it to receive God's message and prepare to be moved by it.

Consider as well Hirschkind's close reading of the self-consciously embodied listening stances described in the Quran. What Hirschkind calls "the physiology of the Quran" rests on three principles.[48] First is the idea that one must listen to sermons not just with the ears but with the heart, specifically with an open heart (*inshira*), as this corresponds to the humble and repentant moral stance that is desired in Islam: God himself, the Quran says, "opened his chest and took the resentment out."[49] Second is the principle that the body is shaped by its own practices, including listening: since "the [perceptual] event occurs within the disciplinary context of a Quranic education, it contributes to the training and inculcation of sensory habits ... in all of [their] kinesthetic and synesthetic dimensions."[50] To "listen with attention" (*al-insat*), as one of Hirschkind's interviewees evokes, is to listen with the entire body.[51] Third, the ideology on which this practice is based includes the belief that language does not exist merely to communicate something about God.

The performative listening, therefore, in which one "follows the ethical movements of the sermon with its appropriate gestural, vocal, and subvocal responses" is akin to a sonic, physiological, and spiritual spooning with God — and, through that intergestural and vocal proximity, to becoming saturated with God.[52] "Ethical listening"—multisensory, highly responsive, and physical—is presented as a material state.[53] Thus the definition of sound is dependent on who is listening. The Quran's ethical listening highlights the listeners' agency in this process: their joint decision about whether to listen fully and, as a consequence, whether to be articulated. What a sound is in such contexts depends on all of the material articulations and circumstances that form the sonorous event and the practices of its participants.[54]

For Hirschkind's listeners as well as for Edison, the crucial aspect of listening is the multisensorial and full-bodied participation of the receiving body, for which Chion's purely aural listening modes cannot account. This idea not only echoes but also expands on the point of Christopher Small's four additional letters—changing *music* to *musicking*[55]—letters that make visible the active doing of what we have traditionally called music. To refer to music as a noun is to align it with objects and thus imply that humans are not necessarily essential to its existence. However, in the words of Brian Massumi, "there's something doing" in music.[56] To turn *music* into a verb is to acknowledge that people, activities, and flows are indispensable to its practice. Furthermore, Small contends that the term *music* reifies an activity—and that it deceptively affirms music (a noun) as a given, a priori, rather than as a reified and naturalized notion. In fact, "there is no such thing as music," Small writes. "This is the trap of reification, and it has been a besetting fault of Western thinking."[57] What the present chapter adds to Small's powerful insight arises from the application of his perspective to the fine-grained level of music making. When we take seriously the idea that music is material and multisensory, we give ourselves the opportunity to pose fundamental questions about music, about what exactly constitutes the object of musical analysis.

We may now revisit Romeo and Juliet, and their analogy with music, one final time. Although Romeo believed that it was Juliet who moved him to love, it was instead their interlocking coarticulations that gave rise to his emotions. Because of his conviction that it was Juliet who engendered his love, Romeo believed that he would never experience other interlocking coarticulations that would give rise to similar feelings. Comparably, we might believe that it is the greatness of an opera that moves us, but it is rather our interlocking coarticulation with the work's musical and dramatic material that bring about our feelings. If we insist on a belief in fixities, we grant ourselves access only to certain people, musics, and sonorities, and we endorse and validate the experiences we have under these now-predetermined labels even if other articulations that we cannot imagine might also offer meaningful experiences. It is this feedback loop—the measurement and comparison of experiences in relation to a roster of sanctified labels—that gradually ossifies our experiences, and that we fix into a particular understanding of music.

The perspective I offer in this chapter attempts to turn the dominant definitions of music, singing, pedagogy, and listening on their heads. Rather than assuming that the emotion of love is evoked by a particular person, or that the transcendent feel of music is evoked by a particular composition, I suggest that these effects arise instead from material coarticulation, and that a variety of human and musical articulations could result in equally powerful effects and affects. Recalling chapter 1's discussion of music's material base, and chapter 3's notion of the naturalization of sound, we may now flesh out and extend their suggestions about the action-based framing of music and voice, and unsettle the conventional understanding of core and supplement. Thus this chapter rejects the idea that sound is the center of music, proposing instead that articulatory action is at music's core. In other words, the idea of the thirteen-year-old girl, Juliet, or the idea of the sound fixed as A# are supplements; the material articulations are the core.

Conclusion: From the Figure of Sound to Material Practice

Singing and listening are not limited to the vocal cords or to the organ we call the ear. Rather, we use the entire body to sing, and the entire body carries out the function we normally locate in the ear. We are all voice, all ears. Beethoven and Edison were each able to reconfigure their material situations to sense sounds that were otherwise unavailable to them. These two listening scenarios poignantly illustrate that sound is not a stable entity that may be captured by a uniform listening practice. Rather, what is sensed relates directly to how sound is listened to, and how the material interactions involved in listening are configured. As I have attempted to show, the material circumstances, the spatiality of the material configurations, and the specific people who collectively participate in unfolding material articulations define what we more conventionally imagine as a fixed song, a piece of music, or a work. Indeed, using a multisensory framework, we understand that voice, song, sound, and music cannot be defined without accounting for the person who senses and the specific circumstances within which she or he transduces. This insight begs us to reassess the roots of music's meaning making. Music's affect and relative value are not limited to any preconceived meaning, ideology, or structure of power such as the work concept. We can begin to consider the ramifications of our setting movement into action through music, and of our being moved by sound. We may realize how and why sound not only has a discursive impact and further implications, but is also of material consequence.

In shifting our understanding from viewing sound as the naturalized primary parameter of music to recognizing music as a material articulatory process, we come to understand that when we make and otherwise participate in music, we do not do so at arm's length. When we make music, we have a material impact on the world. Our musical actions have material consequences. In the moment we coarticulate or reject the music presented to us, we coarticulate or reject part of the person presenting it. In other words, by articulating music—not only as singers, but also as listeners—we are the music. And it is in our being the music, and in our music's becoming a part of others, that music's meaning, power, and consequent ethics lie—because, to quote Benjamin Piekut's succinct formulation, "every musical performance is the performance of a relationship."[58]

5

MUSIC AS A VIBRATIONAL PRACTICE
Singing and Listening as Everything and Nothing

The unnamable is the eternally real.
Naming is the origin
of all particular things.
—LAOZI, *Tao Te Ching*

[The scientist's] religious feeling takes the form of a raptured amazement at the harmony of natural law, which reveals an intelligence of such superiority that, in comparison with it, all the systematic thinking and acting of human beings is an utterly insignificant reflection. This feeling is the guiding principle of [the scientist's] life and work.
—ALBERT EINSTEIN, *The World as I See It*

Stendahl [Marie-Henri Beyle] wrote that music was the highest form of art and that all the other forms really wanted to be music. This was of course a Platonic idea, all the other art forms depict something else, music is the only one that is something in itself, it was absolutely incomparable. But I wanted to be closer to reality, by which I meant physical, concrete reality and for me the visual always came first, also when I was writing and reading, it was what was behind letters that interested me. When I was outdoors, walking, like now, what I saw gave me nothing. Snow was snow, trees were trees. It was only when I saw a picture of snow or of trees that they were endowed with meaning. — KARL OVE KNAUSGÅRD, quoted in Mark Sussman, "Fueled by Sentences: The Uncanny Art of Karl Ove Knausgaard"

At the outset of this book, I set out to try to understand how it is that music can both restore and destroy. More specifically, I asked how it is that what we think of as the same music can have radically different effects, at different times, on a single person. The book took as its premise that investigating music from the sound, signal, and signification (or symbolic) levels cannot fully explain this conundrum. In taking this position I joined a number of scholars who have set out to examine music and musical experience beyond the parameters relied on by music discourse. I have thus far posited that the specific ways we have conceptualized music's building blocks—as species of sound with particular parameters—limits our knowledge of music's ontology, epistemology, and ramifications. When I write that music is connected to sound, I do not necessarily refer to sound as work. Scholars have defined the activity of music, or musicking, with great nuance. However, while much broader practices than those delineated by the work concept are addressed here, an assumption about cultural, performative, or scientific practices involving the "art or science of combining vocal or instrumental sounds," to borrow the definition of music in the third edition of the *Oxford English Dictionary*, is common. Even if we cannot reach unified notions of the "art or science of combining vocal or instrumental sounds" and the practices surrounding that combination, including dance movements (which are soundless, or the sounds they make are not part of their intention), I am not aware of any musical cultures whose concepts of music do not include dealing in some way with notions of sound and silence. By denaturalizing these parameters—indeed, by denaturalizing the notion that music deals in the currency of sound, silence, and chronological time—music can be investigated as nonstatic, not limited to the aural sense and dimension. Instead, we can intentionally investigate music as action, materially transmitted and propagated.

From this fuller view, we have come to the formulation that music is the practice of vibration.[1] Tied to this realization is the understanding that music is neither external nor measurable. From this perspective, we can answer the question that initially motivated the project: since music is always materially and relationally contingent, it is never the same external force that both restores and destroys. Rather, since music is vibration, there are multitudes of material circumstances that contribute to each of its particular articulations, each unrepeatable and hence unique, and each with a potential to affect us that can be revealed only in the particular articulation that takes place within and among each material situation and unique listener.

My initial inability to answer the question arose from my assumption that a single cause was producing different outcomes. An intermaterial vibrational perspective explains how something that, in the figure-of-sound framework, appears to be the same music can lead to very different outcomes for the same person.[2] That is, vibrations that enter my body are transmitted by and through the body at the very point in time that I sense it. More precisely, "it" is not something external that an internal "I" senses. There is indeed no separation between "it" and "I": each configuration forms a unique node and is best understood when investigated as such. The transmission and vibrational configuration are unique and unrepeatable in any dimension. In other words, since sound cannot exist in a vacuum, a given material circumstance and its articulation comprise as much of what we provisionally understand as the sound as what we may point to as the sound or the music.

Ironically, now that we have understood how the so-called same sound can have both restorative and destructive effects, we can also begin to understand that the very question oozes out of a logic that relies on the figure of sound. Indeed, the question's specific articulation must be reconsidered in light of its answer. In other words, the concept evoked—the same sound—implies fixity and knowability, in the same way as does the question "where does the sound begin and end?" There is no "same" music. There is no externally fixed music that is passed on in an unchanged form into vastly different environments, causing different effects in people. There is only the music that comes about in a particular material-vibrational transmission, and the delineation and meaning we give to that transmission. In this way, the music that restores a particular person is, by definition, not the same music that destroys another.[3]

In other words, it is the figure-of-sound framework's ways of understanding music and vocabulary that reduce two drastically different events to what we conceive of as the same event. It was out of the desire to untangle such assumptions that the motivation to write this book arose. The previous framework constrained us to understand music's effects in the world as symbolically based. Due to this assumption, some of music's outcomes and effects seemed illogical, or were unexplainable within that framework. As we have seen, there were no existing analytical frameworks that could deal fully with the idea that the same music could affect the same person with different outcomes.

If sound and music have been reduced to static nouns, then the practice of vibration is a verb—regenerating its energy through material transmission and transduction within a continuous field. (And it is listeners who delineate this continuous field into nodes according to a priori parameters.) In other words, in applying the intermaterial vibrational framework, we turn our focus away from the effect and force of a stable object and toward the unfolding intermaterial process. This refocusing suggests that we need to reexamine not only assumptions about music, but also the ways in which questions are framed and thus what kind of knowledge is sought in that questioning. Moreover, our realization suggests that this shift moves musical inquiry away from the musical objects we study, and toward an intrinsically human and material realm.

To this end, as a result of examining the details of what happens in the doing, in this chapter I propose that an investigation into music is an investigation into relationships and community. Thus, I explicitly move the discussion about music and its ramifications out of the orbit of the knowable and the potentially meaning making, to the material and always already relational.

The Limits of the Symbolic

In which ways may conceptualizing sound and music through the figure of sound—sound and music as knowable—limit not only our understanding of the musical event and our own contribution to it, but also its range of power? Recall from chapter 3 that defining and naming music as distinct sound is akin to nudging a child into the symbolic realm. In the same way that an infant's thick event is usurped and reduced to sound, music's thick event is naturalized and reduced to sound. And in the same way that the sounds "mama" and "papa" are severed from the infant's thick output—"a pure manifestation of vocal resonance linked to a state of internal displeasure"—music as knowable sound is severed from the thick event of musical experience.[4] Over time, what was manifested as a thick event is slowly stylized into a signifier, overlaid with the signifieds *maternal caretaker* and *paternal caretaker*—or sound and silence.

Considering the symbolic in relation to the operatic voice, Michel Poizat notes that the child's cry is not only responded to, but also "*attribute[d]* meaning" and "*interpret[ed]* . . . as a sign of hunger or thirst or whatever."[5] Similarly, the thick event of musical experience is carried into the symbolic order when knowable sound segments are understood within systems of sounds and music and overlaid with a given sound's contextual functionality (say, the second step in a scale) and cultural and historical meaning (say, a commercial jingle). Poizat poetically describes the process of entering into the symbolic: "This first, pure cry is qualified as mythical or hypothetical because as soon as it is interpreted and elicits a reaction, its original 'purity' is lost forever, as it is now caught up within the system of signification that is already in place with the intervention of the Other."[6]

Similarly, in the same way that we generally believe—and act on our belief—that once the symbolic order has opened up for a child, he or she has moved fully from the presymbolic to the symbolic order, so we generally respond to sounds we can name as music, and thus constrain within the symbolic order, with the presumption that they occupy only that order. By defining and naming a sound within the thick event of music, we bring music into the symbolic and subsequently constrain our relationship with it to the limits of the symbolic. But music still operates on the nonsymbolic level. In other words, while people may operate within the symbolic order, the presymbolic does not cease to exist. Similarly, while we can meaningfully understand much music within the symbolic order, music continues to influence us within the presymbolic domain.

For example, in his work on reggae sound systems in Jamaica, Julian Henriques addresses the dynamic between music constrained within the symbolic order and its force within the presymbolic order.[7] Henriques takes up the challenge of describing and explaining the force a community of participants in an event creates and acts upon around the vibes that flow between a given sound system, music, and the community. While the forces described are not constrained to the figure of sound and the symbolic, the discourse around them is.

Henriques's strategy is to carefully map the technical aspects of the sound system, and humans as material and cultural agents. He concludes with a notion he terms "*sonic logos*." With this concept, he seeks to capture the "kind of reason that emerges from the relationship between two distinct senses of sound."[8] With the idea of the sonic logos, then, Henriques seeks to encapsulate the "practical activities of soundings," as in "the periodic movement of compression waves through a material medium," with "sociocultural *values*." Henriques contrasts his approach with the way in which, in a sociocultural context, we can "[get] a sense of" something by seeking to "know" as much as possible about it beyond its nonmeasurable qualities.[9] With the sonic logos, Henriques explains, we move to, from, and between the body of sound and its mindfulness—that is, between how the sonic logos might be "'built' from the crew's ways of knowing and skilled techniques to understand[ing] how they 'make sense' of what they do. It thus explores the *relationship*, connection and intertwining of body and mind together."[10]

Henriques's work exemplifies one strategy used in attempts to understand dynamics beyond the symbolic. Sound's affect beyond the symbolic and logos is captured by Emma Dillon's phrase "sense of sound," while the recuperation of vocality through thirteenth-century French manuscripts and Shane Butler's tracing of the writer's hand and voice through written words and the materiality of book and handwriting have also been influential to my thinking on this matter.[11] Furthermore, gravitation toward rationally explaining a fixed and defined musical phenomenon, often dealt with through the application of hermeneutics despite the felt necessity that music reaches beyond its bounds and promise, is encapsulated in Phil Ford's repurposing of Christopher Lehrich's phrase "magical hermeneutics."[12]

It is this tension, so poignantly captured by Ford, that Poizat examines in his work on opera fandom, the fans themselves, and their desire for operatic voices.[13] For Poizat, coming from a psychoanalytical perspective, the thrill of consuming an operatic voice is found in its "failure of speech and the signifying order."[14] He also refers to this condition as the "pure cry," which exists "outside musical discourse" as "sheer vocal effect."[15] He offers a historiographical speculation in which opera is read as a slow progression from speech-like song, within which "singing grows more and more detached from speech and tends more and more toward the high notes; and culminates in the pure cry." He defines this "pure feeling" that comes from listening to music as not "mere pleasure," but rather as "jouissance."[16]

In a related view, David Schwarz posits that "musical listening subjects are produced when moments in performed music allow access to psychological events that are presymbolic—that is, from that phase in our development *before* our mastery of language."[17] Thus, "music can remind us of phases in our development when we crossed from imaginary to symbolic experience, and that the musical representation of such threshold-crossing produces listening subjects."[18] Indeed, for Schwarz, we "listen to music as an attempt within the symbolic order to hear echoes" "of presymbolic experience we hear into the spaces between symbolic convention and presymbolic sound."[19] However, again we find ourselves pitting the sensible and symbolic against the nonsensical and nonsymbolic.

Poizat, however—like Henriques, Dillon, Butler, and Ford—seeks to avoid this separation by embarking on a discussion of the materiality of the voice. According to Poizat, we are always in search of the "vocal materiality" and the "simple phonic materiality" that we experienced before the voice became symbolic.[20] On the one hand, he argues that it is this "lost object" that the opera fan is trying to regain. But, on the other hand, he also says that it is not truly a loss—in his words, "there is merely a 'loss effect.'"[21] Because the subject has "unconsciously made it the representation of a totally purified trans-verbal state," the presymbolic is understood as lost.[22] But "the elation felt by the fan, listener, and spectator of opera in those rare and fleeting moments when the music lover is irresistibly captivated" is indeed the work of the present presymbolic "vocal materiality" or the "presumed primitive encounter with jouissance."[23] Thus, voice's sound and materiality constitute both the naturalized mode that prevents access to the presymbolic, and the only portal that can offer entry to it.

In her sociological work on music in everyday life and music therapy, Tia DeNora has examined questions about the body's materiality in relation to music and the role of signification in that process.[24] However, rather than thinking about music and its sonorous link to the body, she suggests that we consider "what music may achieve, silently." More specifically, DeNora posits that while music's "link to the body" is "distinct from music's interpretive processing," the process is not a regression to "a nondialectical understanding of music-as-stimulus."²⁵ Also addressing the traps that can arise when we follow the trail of bread crumbs scattered by the sound of music and its possible function as sign and signified, Gary Tomlinson has noted: "Humans are symbol-makers too, a feature tightly bound up with language, not so tightly with music. . . . Homo symbolicus cannot help but tangle musicking in webs of symbolic thought and expression, habitually making it a component of behavioral complexes that form such expression. But in fundamental features musicking is neither languagelike nor symbollike."²⁶ For Tomlinson, studying humans in relation to the symbolic offers clues about music's ancient origins. And even for scholars without similar ambitions, such discussions offer clues regarding the relationship between the symbolic and music making. As Tomlinson notes, music in itself is not primarily symbolic; rather, it is made to serve the symbolic. If we can release that perspective, he continues, we can "plot . . . the counterpoint between musicking and the language and symbolic cognition that coalesced alongside it."²⁷

One part of the presymbolic phenomenon that some psychoanalytically oriented scholars address is, for me, most clearly explained by Tomlinson. He concludes "Evolutionary Studies in the Humanities: The Case of Music" by pointing out that part of the problem with understanding music is the limited timescale within which we think about it. If we go far enough back, Tomlinson observes, we can clearly see that humans engaged in "musicking" prior to the appearance of "either language or symbols."²⁸ Tomlinson's preliminary million-year evolutionary history of hominids shows us that musicking is neither dependent on, nor an evolutionary outcome of, symbol making. What Poizat understood as the "presymbolic," DeNora recognized as "silent practice," and Tomlinson thought of as musicking independent of language or symbols, I describe as the thick event of music and, ultimately, as intermaterial vibrational practices and events.²⁹ Thus, on some level, while we primarily conceptualize—and, accordingly, have access to these events—on linguistic and cultural planes, we still react perceptually and instinctually to them. If you will, we are affected by the falling of the tree rather than by the sound of the tree-falling event, as indeed our presymbolic and prelinguistic ancestors most likely experienced the phenomenon.

Toward an Organological Inquiry
into Intermaterial Vibrational Practice

Some scholars might react to the proposal that we investigate music from an intermaterial vibrational perspective much as I initially reacted to Juliana Snapper's underwater opera. Recall that I first dismissed her work by asking why I should bother with something so unlikely and cumbersome. In the same way, scholars and others might rightly ask why we should insist on viewing music as intermaterial vibration when, on the one hand (even if we agree that the figure-of-sound framework is incomplete), it works well enough and, on the other hand, in the larger scheme of things, why should we bother with the return to vibration? Is this not such a small detail that hardly anyone would notice? Is this particular emphasis not just a minute shift in perspective? Furthermore, will we not, in reality, continue to think about sound in the way we have always thought about it? In other words, what useful outcome could this perspectival shift have? In response to these objections and questions, I argue that even a cursory examination of music itself as vibrational material matter suggests a few points at which our understanding of music departs from how we understand it through the figure of sound and the symbolic.

First, when music is predominantly understood not only through the symbolic but as material and, indeed, intermaterial vibration, we can begin to conceive of it in a class with—instead of separated from—other intermaterial vibrational phenomena. While John Cage famously included so-called silence in his compositional material, thus questioning the nature of sounds allowed in music and the position of intentionality in their production, understanding music on a par with other intermaterial vibrational phenomena questions the division between material modes, including the exclusion of some from our attention. If we understand our bones and flesh as participating in forming the music we experience, are they not as much a part of the music as the so-called musical work? And if music is the fluctuation and transmission of energy, does it not have something in common with other forms of energy fluctuation and transmission?

Second, when we examine music as vibrations, we see that the object of study is not only the vibrations but also, for all practical purposes, the material that vibrates. Expanding our perspective in this way reveals that the vibrations themselves are shadow phenomena and that, in fact, vibration does not exist prior to a specific material realization. In other words, we cannot experience and define the boundaries of a particular vibrational occurrence before it comes into existence. Indeed, this node does not exist, or is not realized, except at the precise moment when the engaged material vibrates. A familiar image that may offer a useful analogy is the invisibility of the wind, except when it moves a tree branch.

While the musician and philosopher Sun Ra, for example, understood music as the privileged engagement of vibration within a broader cosmology, I came to this place of inquiry through two simple steps.[30] The first was to take seriously what I understood as grave inconsistencies in representations of music's meaning and culture. Taking those inconsistencies and their ramifications—such as why and how the same music could restore and destroy the same person—seriously, I looked into the underlying questions and assumptions. Through doing this, I understood that it was the questions that produced these types of answers. In other words, if the knowledge produced seemed entirely inadequate or irresponsibly incomplete, the second step was to change the questions. So, when I stopped banging my head against the question of the meaning and effect produced by music, my thinking was dislodged enough to enable me to further understand the dynamic of relationality.

In examining the fundamental questions underlying most inquiries about music (which are based on the assumption that music is knowable), I began to understand how theories built on articulated or tacit assumptions about the figure of sound often do not make room for inquiry of and through practice. For example, while the figure of sound promises to explain and systematize the intricacies of music, our dependency on its conceptual framework locks us in perceptually and intellectually, eroding our ability to approach musical phenomena beyond sound. As a consequence, we allow music to be ossified into what initially promised to be merely a useful theoretical framework. The theory of the figure of sound, then, rejects the phenomenon it sets out to investigate, replacing it rather than merely moving us toward partial understanding. As a result, the figure-of-sound theory is woven into the world so strongly that we hardly know when and how we are using it, or when we are not.

Throughout most of this book I have described the way the figure of sound functions, how we can let it go, and what can happen if we do. In the end, we are left with the realization that when we do let the figure of sound go, we become able to interact with music in a way that helps make us better able to answer the question to which the figure of sound purported to lead, but did not: how is it that music does what it does? The punch line is this: the question is not answered theoretically, with reference to the a priori; rather, it is answered practically and with no a priori as a baseline reference. Before we shift to the question behind the question, and before we work through these issues in practice, we have no way of articulating that underlying question. Without breaking through the foundational position that the figure of sound has come to occupy, which prevents us from comprehending music more fully, this question is inaccessible. I propose that we can understand more about music's many dimensions if we do not limit ourselves to the logic presented by the figure of sound, but instead use the perspective of performance and practice to burrow down into expanding notions of music to the material, vibrational, and energetic arenas.

With *Sensing Sound*, then, I suggest that a return to the study of music as intermaterial vibration can afford a better understanding of the ways in which music does what it does, and the ways in which humans use it as a force for good and bad. I propose that this study of listening, singing, and other music making is akin to an organology of intermaterial vibration. A practice that began in the seventeenth century, organology is the study of musical instruments in regards to aspects ranging from their historical aspects and social function to instrument building (design, material, and construction) and performance.[31] In *Sensing Sound*, I assume that the "instrument" is the field of intermaterial vibration, and I have studied the construction of the vocal instrument in relation to, and as constructed through, performance practice.[32]

Thinking about music beyond that which is sounded is, of course, not new. Renaissance thinkers saw *musica*—that is, the study of music within the liberal arts—as "the science of comparing numbers, which could be manifested in myriad external forms."[33] Considering sound as vibration is also, of course, unoriginal. Galileo Galilei's *Dialogues Concerning Two New Sciences* (1638) offered the first systematic study of vibrating strings, and his explanation of the origin of consonance and dissonance remains generally accepted today.[34] And in 1660 Robert Boyle's classic experiment in which a ticking watch was placed in a partially evacuated glass chamber showed that sound exists only when mediated, or transduced, by matter.[35]

"The words instrument and organ are in Latin and Romance languages intertwined," Bonnie Gordon reminds us. "Organ," she explains, "derives from the Latin organum, which means instrument, and which comes from the Greek word that means musical instrument or organ of the body."[36] The problem of the boundaries between "organs endowed with sense and those without" was considered by many people. For Francis Bacon, Galileo's contemporary, there was no difference between humans and mechanical instruments, but organological studies do separate the human from the nonhuman organ.[37] While there is no consensus that such a division is relevant, the relatively recent material turn can help mend that historical gap and habilitate the study of music's materiality. Indeed, building on Bill Brown's thinking about things, by putting into perspective notions of what an instrument is and what an organ is, I examine the collective intermaterial vibrations of music, sound, and voice to ask how music "as an inanimate object enables human subjects (individually and collectively) to form and transform themselves."[38] Focused on vibrating technology in the period when it was recently introduced to the public, Shelley Trower's work shares Brown's interest in material cultures. Examining the culture of the Romantic harp; vibratory objects including strings and spiritualist objects; and the introduction of new technology such as the bicycle, railway, sewing machine, and vibrating medical devices, Trower is concerned with material and literary cultures and objects as "vibratory movements of various material objects" and how "vibration is also bound up with different kinds of materiality beyond objects, including the air and ether."[39] In this way, Trower provides a useful model for historical work on vibration and how it has been conceived, harnessed, and used.

I take the position that studying voice, sound, and music from the point of view of materiality generally, and organology specifically, offers access to an otherwise unrecuperable history and process.[40] That is, I assume that investigating the intermaterial vibration of singing and listening can become a music-analytical operation itself. Studying intermaterial vibration as organology puts voice and perception into conversation with Emily Dolan's observation that "in their ubiquity and diversity, instruments might be thought of as boundary objects."[41] Building on the work of Trevor Pinch and Frank Trocco, Dolan quotes Susan Leigh Star's and James Griesemer's work on boundary objects within museum and exhibition contexts. Star and Greisinger explained that an object's "boundary nature" can be understood as "simultaneously concrete and abstract, specific and general, conventionalized and customized. They are often internally heterogeneous."[42] Bringing the discussion back to musical instruments, for example, Dolan notes that the synthesizer's boundary nature is described by Pinch and Trocco. That is, the idea of the "liminal entity" is invoked to convey not only the crossing of boundaries, but also the transformation that takes place through that crossing.[43]

We may consider material and vibration as instruments, while what are more traditionally conceived of as musical instruments may be considered as subclasses of material and vibration. If we use this viewpoint, our conception of the liminal space and arena necessarily shifts. Voice, with its ability to filter out certain frequencies by changing shape, is often used to describe the synthesizing process. Voice also plays multiple roles as an anatomical entity that protects the lungs from food and liquids, as a sound shaper, and as a transmitter of music and words. With its many roles, voice represents a prime example of the physicality of the "boundary object" or "liminal entity."[44]

As noted throughout this book, some of the material that vibrates during a musical experience is the human body. By considering music in this way, we may study the phenomenon of the "boundary object" or "liminal entity" through observing shifts in vibrational patterns. We may also come to terms with the notion that we are one and the same as the vibrations. Rather than viewing music as an external and stable object, signal, or ground for meaning making, it is these various intermaterial vibrational states (thick events) that we transmit and take in, that we interpret and make meaning with, and that we refer to as music. Thinking through intermaterial vibrations, we learn that we are putting ourselves on the line.

However, by pursuing intermaterial vibration within a music-analytical context, I have no illusions that I am adding to the existing technical knowledge about these topics. Because much of current music discourse has moved so far from directly drawing on this knowledge (although the knowledge is already present in related fields), my modest hope is to merely reengage some of the terms. My point is precisely that this book does not offer new insights. Instead, it engages insights arising from the observation that music discourse limits its understanding of music to sound, silence, and the practices that surround them. Recall from the introductory chapter that I position vibration (as in sound), transmission (as in intermaterial flow), and transduction (as in the conversion of wave forms from, say, mechanical to electric) within historical and theoretical discourses. I draw on this rich body of knowledge to distill my proposition that analyzing music, and life in music, from this perspective can tremendously enrich our understanding of how music does what it does.

Intermaterial Vibration and Energy

In the following visual representation created by R. Bruce Lindsay explaining vibration, the severance of sound and music from other broad areas of material vibration, transmission, and energy is telling. As I noted earlier, when we believe that music deals in the figure of sound—the currency of sound and silence—implicit or explicit questions nudge us into identifying its distinct units (types, qualities, and temporal span) and form (From which units is music built? When does the music stop and end?). In figure 5.1 we can see that music is largely conceptualized as constrained by the limitations of musical scales and instruments, with a much smaller overlap with areas such as communication and room and theater acoustics. While some might differ re-

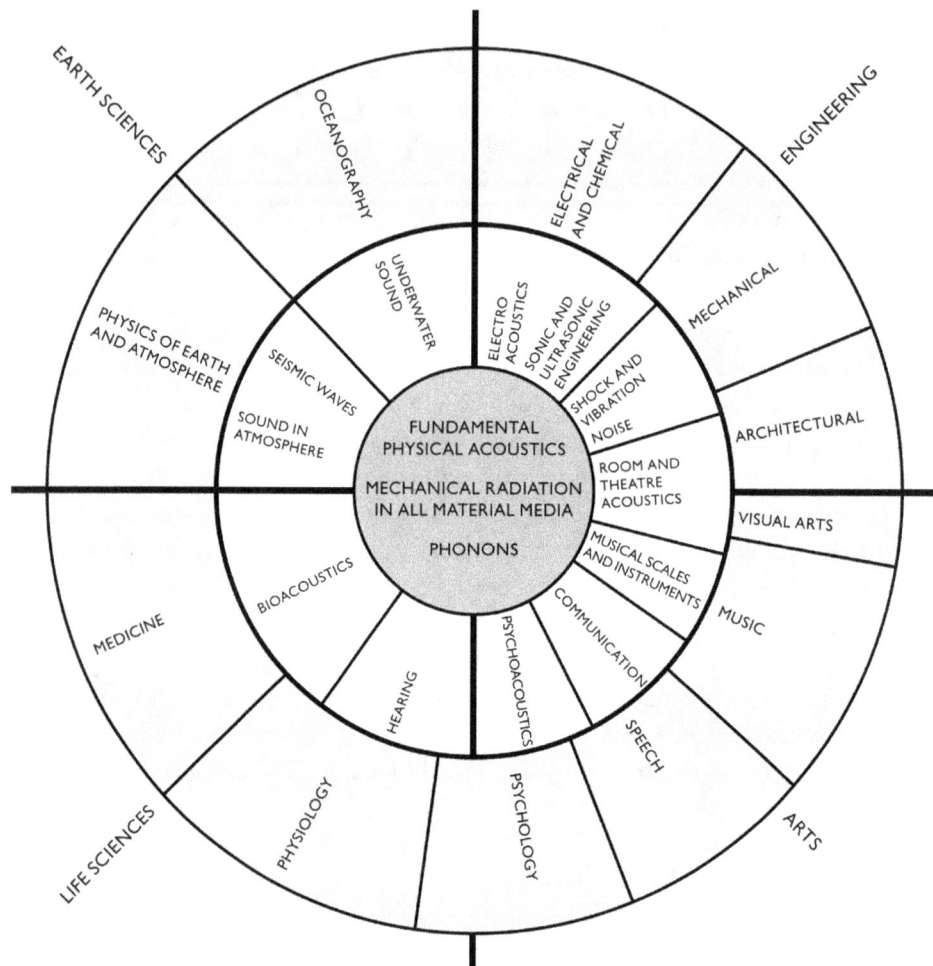

FIGURE 5.1 · This figure has become known as the "Lindsay's Wheel of Acoustics" (R. Bruce Lindsay, "Report to the National Science Foundation on Conference on Education in Acoustics," *Journal of the Acoustical Society of America* 36 [1964]: 2242).

garding the specifics of the chart's layout (for example, some might say that psychoacoustics also falls within the category of music, rather than only within psychology and speech as the chart indicates), I believe the chart outlines the general practice of the scholars in the fields indicated.

In summary, the figure shows the division of music's ontology and epistemology from the other areas of intermaterial vibration. Thinking about voice, listening, sound, and music from an organological point of view can, while making use of a traditional perspective, lift the scrim that divides music from other related areas of inquiry. Once we are aware of this division we can also appreciate its randomness, develop an awareness of the connections, the definition of the knowledge area, the practices of the inquiry, and the necessary insights gleaned, and engage with them. We can then use that knowledge to understand more about how vibration operates on us, drawing on knowledge in areas that are connected through what the chart visually presents as its core: fundamental physical acoustics.

From the point of view of vibration, human entities constitute only one of the many materialities through which energy is transmitted and transduced. It is the full spectrum of transmission and transduction of sound vibration that is taken into account in explicit studies of the impact of vibration in positive (therapy) and negative (regulations regarding health hazards) terms.[45] However, these areas of vibrational inquiry are often bracketed off from the study of music and its aesthetic, social, historical, and cultural considerations. As in the limiting of musical study to the examination of only certain vibrational frequencies, I believe the exclusion of what is commonly considered the non- or lower-aesthetic applications and impacts of vibration—for example, music therapy—leaves the study of music per se and our knowledge about its power incomplete.[46]

Douglas Kahn has also observed the bracketing of sounds or energies that were granted "musical or aesthetic status."[47] Specifically, the two different modes to which he refers are those created by wind and natural electromagnetic activities. Kahn finds it ironic that Henry David Thoreau's description of sounds produced by telegraph lines vibrating in the wind[48]—vibrating strings have been aestheticized since antiquity when heard in nature or from constructed instruments, including the Aeolian harp—are taxonomized as a piece of music. The irony arises because the sounds produced by telecommunication lines ("natural radio," heard by Thomas Watson, when, as Kahn describes it, "the long iron telephone text line acted unwittingly as a long-wave antenna"[49]) were not aestheticized until much later. Kahn asks why the sounds "created by the wind [were] granted musical and aesthetic status through the category of the Aeolian, while the sounds created by the natural electromagnetic activity were not, even though they were heard musically and aesthetically, could occur on the same line, and were produced in the same environment?"[50] In many aspects, my question about sound's epistemological status is not unlike Kahn's. I might rephrase his question along these lines: What grants certain aspects of energy musical and aesthetic status, how are knowledge and practice formed as a consequence, and what effect does this have on our music making and listening?

Again, if we were to accept an organology of intermaterial matter, we would also need to critically examine the genealogies of various academic fields and their divisions into new areas of inquiry. For example, in a 2007 text, Singiresu S. Rao — a mechanical and aerospace engineer who wrote one of engineering's basic textbooks — traces his profession to the general topic of vibration, which formerly united fields that today are entirely separate.[51] In *Vibration of Continuous Systems*, he locates the origin of engineering in "the earliest human interest in the study of vibration," specifically in the form of experimentation with musical instruments.[52] Humans, Rao writes, have "applied ingenuity and critical investigation to study the phenomenon of vibration and its relation to sound."[53] In a sweeping historical overview of vibrational study, Rao writes about key characters in the history of vibrational study fundamental to mechanical and aerospace engineering.[54]

The physicist R. Bruce Lindsay, creator of what has become known as the "Lindsay's Wheel of Acoustics" shown above (figure 5.1) and a respected member of the Acoustical Society of America, was much concerned with the historical and philosophical aspects of energy and entropy. In a 1966 article, he offered a historical overview of the "problems of acoustics" according to the field's tripartite division: the production, propagation, and reception of sound.[55] After tracing the history of these areas, Lindsay ends his overview with a cautionary note to some of his colleagues, those "physicists who are carried away by the glamor of high-energy physics and the properties of the solid state." While these physicists may think that "the future of a so-called 'classical' field of physics like acoustics lies wholly in its technological applications and that as physical science it is 'played out' . . . there is no ground for assuming that man will ever run out of questions about acoustics any more than he will run out of questions about the nucleus and the theoretical particles that inhabit it or can be created from it," Lindsay boldly claims. "What is, of course, true," he underlines, "is that as investigation proceeds, the boundary lines between the various types of natural phenomena that mankind has artificially erected for purposes of convenience are becoming fuzzier and more unrealistic. The aim of the science of the future is a meaningful synthesis."[56] Lindsay takes this point from the history of acoustics, an extremely interdisciplinary endeavor in which each area of investigation would be diminished if it were not understood as intimately connected to the others.

Rao's and Lindsay's examples of the origins of acoustic and vibration theory are wide-ranging: from stringed instruments in China, India, Japan, and Egypt from 4000–3000 BC to Pythagoras's study of vibrating strings in the sixth century BC; from the seismographic earthquake measurement by the Chinese historian and astronomer Zhang Heng in AD 132 to Galileo's work in the sixteenth and seventeenth centuries—including his measurements taken with a simple pendulum and vibrating strings, which are considered to have laid the foundation for modern experimental science; and from Isaac Newton, whose differential calculus regarding the laws of motion was later applied to previously unsolved problems in mechanics and physics, to Sophie Germain's work on the vibration of plates and shells and Lord Rayleigh's theory on sound and vibration.[57] However, despite millennia of inquiry into vibration, Rao notes that it was only around forty years ago that "vibration analyses of even the most complex engineering systems were conducted using only a few degrees of freedom"—made possible thanks to the 1950s advent of digital computers.[58] As we can see, new applications for the study of vibration continue to be introduced, indicating that there is much more knowledge to be gained.

Throughout this book's first three chapters, I sought to denaturalize the notions on which we depend in order to consider music: music's transmission through air, music as independent of space, and music as trading in sound.[59] In our interest in understanding more about music's complex power, we continue millennia-long investigations into music as vibration. The question that follows is: how may we think organologically about intermaterial vibration, transmission and transduction in general, and spatial specificity and sound specifically?

From the Figure of Sound to Perspectives on Vibration, Transmission, and Transduction

Let us think about vibration, transmission, and transduction by expanding on the foundation that, as an example, twelfth-grade students in the United States are required to know. The National Research Council Committee's standards recommend that by the end of twelfth grade, students know that "the wavelength and frequency of a wave are related to one another by the speed of travel of the wave, which depends on the type of wave and the medium through which it is passing. The reflection, refraction, and transmission of waves at an interface between two media can be modeled on the basis of these properties."[60] This definition of "wavelength and frequency" can be translated for our purposes by relating it to hearing. In these terms, hearing really means to materially participate in the "reflection, refraction, and transmission of waves"—which can also be imagined simply as energy transference.

Energy states' movement across the material with which we are concerned can be divided into two major classes, transmission and transduction. Transmission describes vibration taking "place through an elastic medium by means of wave motion" and refers to vibration or energy that is contained—in other words, mechanical energy in one state remains as mechanical energy in the transformed state, and there is no change in the type of wave motion.[61] Points of transmission include all nodes—for example, the vibrating guitar string, vibrating air inside the guitar's body, the air outside the guitar, and eardrum—of the phenomena that we might think of as guitar music. Transduction describes energy that is transformed from one state to another, such as from mechanical to electromagnetic energy. For example, a transduction takes place during the process of translation from the eardrum's mechanical vibration from the guitar to the electromagnetic signal in the nervous system.

Using our example of the guitar, we can also modify our example to include a transduction that takes place earlier. If we are using an electric guitar, say, the transduction takes place as the signal passes from a vibrating string to the electromagnetic pickup that receives the signal and transmits it through the amplification system and thence to the air again. In the case of hearing, transduction takes place when the vibrational energy is converted from mechanical to electrical—that is, when a sound is transmitted to the air and detected by the human hearing organ—specifically, the organ of Corti, where hair cells serve to transform mechanical vibrations into neural impulses.[62]

Moving beyond the basics that U.S. twelfth graders are required to know, the mechanical engineer Chandramohan Sujatha explains that the study of acoustics originated in the "study of vibrations and the radiation of these vibrations as acoustic waves. [Acoustics] deals with all aspects of production, propagation, control, transmission, reception and effects of sound. These are applicable to sounds created and received by human beings, machines and measuring instruments." Today, acoustics includes the study of "sound and mechanical waves in gases, liquids and solids." Sound available to the human ear, Sujatha writes, "is defined as any pressure variation over and above the mean atmospheric pressure impinging on the ear drum; it encompasses all sounds that the ear can detect, from the weakest sounds which are barely audible to sounds which cause pain and damaged hearing."[63] (According to conventional audiology, the young, healthy human ear is capable of detecting frequencies in the range of 20 hertz to 20 kilohertz.)

While we privilege those vibrations that are detectable by the naked ear, throughout this book I have argued for the ways in which the production, propagation, control, transmission, reception, and effects of what we do not currently name in music discourse powerfully participate in and shape our overall musical experiences. It might be these infrasound (below 20 hertz) and ultrasound dimensions (above 20 kilohertz), dealt with in acoustical studies not concerned with music, toward which DeNora gestures in her term "musicking as a 'Silent' Practice."[64] It may also be that these vibrational modes are not transmitted through the eardrum or the inner ear, but throughout other areas of the human body. Furthermore, while acousticians distinguish vibration as motion through solid material (viewing acoustic waves as disturbances of the air), we sense vibration indirectly, through the air motion induced by the vibration. And this air motion can elicit vibrations in a solid mass.

If we move beyond the field of music to examine wider scholarship on vibration, we learn that the "human reaction to vibration is a function of amplitude and frequency of acceleration applied to the body, direction (vertical and horizontal) and character of the motion (linear or rotation)."[65] We also learn that while the human body overall has a low-frequency resonance, it does not vibrate as a single mass. Indeed, beyond the ear, different parts of the human body have their own natural frequencies. As a consequence, individual body parts resonate with the overall vibration—resonance that leads to amplification or attenuation of various vibrations in particular areas of the body. Clearly, expanding our inquiry into listening based on the figure of sound to include the understanding of intermaterial vibration as "amplitude and frequency of acceleration applied to the body" allows us to more thoroughly explore the possible material ramifications of music—and thus we may also expand our understanding of the meanings we make with music. I turn now to a few preliminary suggestions regarding where areas that were previously naturalized through the figure of sound can be expanded.

In chapter 1, I posited that sound is materially transmitted and that music is uniquely and unrepeatably realized in each material node. Because discussions about music have been limited to the logic of the figure of sound, this phenomenon has primarily been discussed in relation to areas of resonance and loudness. Thinking organologically, each body or object, even if it appears to be wholly inanimate, is constantly in motion at the molecular level. By the same token, every object vibrates, quite naturally, at certain frequencies.[66] Resonance refers to the coincidence of two phenomena: for example, a glass resonates because (1) an external sound, vibration, or other force—such as the soprano's high note—matches (2) one of the glass's natural modes of vibration, causing it to vibrate vigorously at exactly that frequency. The story of the soprano who hits a certain high note and shatters a crystal glass may be explained as a phenomenon of resonance: the frequency (tone) of her loud high note coincides with the frequency of a natural mode of vibration inherent in the glass, thereby inducing a vibration vigorous enough to break it.

However, in an organological investigation of intermaterial vibration, the nodes we think of as sound can be investigated as nodes of transmission. And the investigation can be expanded beyond musical instruments into every node that is affected during the experience of music. An organological investigation of intermaterial vibration would consider music in its vibrational realization, in how it is realized by different parts of our bodies, and in how the combination of material nodes reconfigures another node into transmitting or transducing energy uniquely. The boundary drawn around the object, or the liminal space that appears as the boundary is drawn, is reconceptualized from the sound's volume (the perspective of the figure of sound) to a transformation on the molecular level (the perspective of the organology of vibrations) when we consider the shattering of the glass.

For example, the abdomen is highly sensitive to vibration.[67] In the abdomen, resonance in the vertical direction occurs in the 4–8 hertz range and can amplify a vibration up to 200 percent. The neck and lumbar vertebrae amplify vibrations of 2.5–5 hertz up to 240 percent. Certain vibrations even set up a strong resonance between body parts. For example, 20–30 hertz vibrations amplify the head-shoulder resonance by up to 350 percent. Certain frequency regions are especially resonant with organs or body parts that perform specific functions. Vision can be affected due to the 20–90 hertz correlation with the resonance of the eyeball. Sujatha's vibratory model of the human body is reproduced in figure 5.2. In addition to taking into account how vibrations affect distinct regions of the body differently, it is also necessary to consider that the body's sensitivity to vibration is affected by its general posture (for example, whether it is standing, lying, or sitting).[68]

Excitation of a vibrating system can take the form of displacement and/or velocity (the rate of positional change) of the mass element or elements. The change imparts potential and/or kinetic energy to the system. Due to the initial excitation, the system is set into oscillatory motion, which can be called *free vibration*. It is during this free vibration that an exchange takes place between potential and kinetic energies. Within a conservative system, the sum of

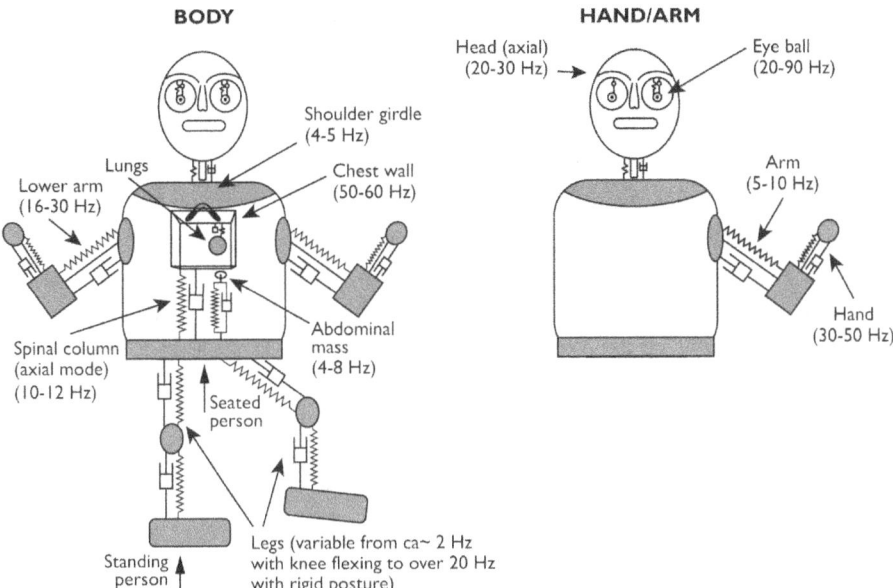

FIGURE 5.2 · "Vibratory Model of the Human Body" adapted from Chandramohan Sujatha, *Vibration and Acoustics: Measurement and Signal Analysis* (New Delhi: Tata McGraw Hill Education, 2010), 295.

potential energy and kinetic energy is constant at any point in time. Theoretically, the system continues to vibrate, but in practice the surrounding medium (for example, air) causes dampening or friction, and thus energy loss takes place during motion. Therefore, for the system's vibration to be maintained in a steady state, the energy that dissipates due to damping must be continually replaced. This means that a vibrational impulse is not stable or independent of material circumstances. Its excitement depends on the general vibrational condition (say, in an enclosed area versus in the open air), the concentration of masses (such as bodies), and other material characteristics.

Among other spatial and relational investigations, an organological inquiry into intermaterial vibration could explore the dynamic between the body's sensitivity to vibration and how the body is affected by oscillatory motion or reconfigured by the extended material vibrating continuum of which it is part. This dynamic not only raises questions about the body's response to vibration; it also asks how this process works within a given material configuration, and how that response affects the body's expression. Because impedance is mismatched, transferring energy from air to the human body is not as easy as transferring energy from mass to mass. Like my examinations of Snapper's underwater opera project, further examinations of the body's particular material configuration and relation to mass in the acts of making and listening to music could lead to new insights into musical experience, as well as suggest specific material configurations and interactions with music.

In chapter 2, I discussed the always already spatial aspect of sound. For most animals, including humans, the ability to localize sound is due to the placement of the two ears as far apart on the head or body as possible. This placement, and additional information gathered as the head moves to collect more data, results in the ability to hear in stereo and locate sounds in space. Because the two ears are situated differently in relation to the sound source, we draw on two distinct sources of information to compute sound's spatial relation to us. Its location is understood in terms of its three-dimensional position and velocity.[69] Additionally, we understand sounds within enclosed spaces by distinguishing direct from reflected sonic signals. That is, we compute the difference between the original sound source and sounds emitted from other locations. This perceptual procedure, referred to as the precedence effect, allows us to automatically localize a sound to, say, a person's moving mouth, rather than to other surfaces that reflect the sound.[70] Thus, Jens Blauert writes, "the totality of all possible positions of auditory events constitutes auditory space." He concludes: "The word 'space' used in this expression is to be understood in the mathematical sense, as a set of points between which distance can be defined."[71]

Not only is sounded sound always already spatially and relationally specific (and thus would benefit from being studied as such), but our perception of this specificity is complex. Therefore, thinking through the issue of sound's specific spatial placement and relationality to the listener, or of receiving vibrational nodes, organologically can provide some tools that may help us sort out the issue's complexity. Even when we become aware of sound's spatial specificity, that localization is not a straightforward matter. For example, in outlining the major principles of the "psychophysical territories of spatial hearing," Blauert shows that different sensory organs (including the presence of one or two ears, vision, balance, touch, and reception of tension) participate in the process of locating sound, and that a number of different notions beyond monaural and binaural theories for air-conducted sound must be engaged to interrogate these phenomena (including bone, visual, vestibular, tactile, and motional conduction theories).[72] Auditory events may occur in any directional relation to the person who senses the sound: inside his or her body or behind objects, near or far. Additionally, factors such as familiarity with the sound and the specificity of human anatomy play a role in identifying and locating auditory events.[73]

Keeping these nuances in mind, Blauert's terms "locatedness" and "localization" can be useful in looking more deeply into spatial specificity and relationality. He uses the term "locatedness" to describe "the spatial distinction" of a sound: "The locatedness of an auditory event is described in terms of its position and extent, as evaluated in comparison with the positions and extents of other objects of perception, which might be other auditory events or the objects of other senses—in particular visual objects."[74] In short, the concept of "locatedness" facilitates a discussion that takes into account the complexity of sound as it is sensed within space. For instance, while the position of a sustained sound in a reverberant room is not easily pinpointed, a short sound in an anechoic chamber can be precisely located.

"Localization," Blauert explains, "is the law or rule by which the location of an auditory event (e.g., its direction or distance) is related to a specific attribute or attributes of a sound event [physical sound source], or of another event that is in some way correlated with the auditory event [in which the sound sounds like it is emitted]."[75] Localizing a sound seems pretty straightforward. However, the *sound event*, the *physical sound source*, and the *auditory event* are not identical. For example, under certain circumstances, the same sound event can yield simultaneous yet differing auditory impressions. Furthermore, Blauert reports, "localization varies within certain limits from one subject to another and undergoes nondeterminable variations over time."[76]

Among other spatial and relational investigations, an organological inquiry into intermaterial vibration could examine the dynamic between localization and locatedness. These terms partially explain the complex nuances of technical descriptions of spatial specificity, its psychoacoustic aspects, and the relationship between sound sources and sounds reflected from reverberating surfaces. The terms also offer just one example of a possible further organological direction of inquiry into both the spatial specificity of the musical event and the way in which the meaning we form around the vibrational impulse is partially based on our spatial sense of it.

In chapter 3, I discussed the naturalization of sound as one of music's currencies and suggested that what we think of as sound might be better understood if we conceive of it as vibration. However, an organological investigation into the matter would quickly reveal that the phenomenon is more complex than this. That is, while I posit that what we have identified as sound is a species of vibration and would be more accurately and usefully described as such, the twist is that while music is not necessarily sound, not all sound is related to a causal event. In other words, sound signals and auditory events are not necessarily related, and an auditory event has not necessarily been preceded or caused by corresponding mechanical vibrations or waves.

Thus, an organological inquiry into the naturalization of sound would, first, investigate the nuances of the phenomenon that we understand as sound—beyond the confines of the assumption that sound is caused by intermaterial vibration. Some auditory events that are not caused by a sound signal but by other factors could include auditory hallucinations, disease conditions (such as tinnitus), or sound experienced as the result of artificial stimulation of the acoustic nerve.[77] Our organological inquiry could also encompass ranges of vibration including "'mechanical vibrations and waves of an elastic medium, particularly in the frequency range of human hearing (16 Hz to 20 kHz)."[78] Moreover, any investigation must be mindful of the way in which, as Blauert observes, these "physically measurable changes of position" are based on what is primarily perceived visually, and of the fact that the range of vibrations and waves with which we are concerned are limited to those that fall within the "frequency range of human hearing."[79] In other words, an organological approach to the question of sound would inquire into the limitations and ramifications of certain causes of sound (that is, why traditionally priority is given to inquiry into shared or sharable experiences instead of hallucinations?) and into the privileging of certain modes of experience (human over other animals) and nodes of materiality as the basis for vibration (human entity over object).[80]

In chapter 4, I posited that what could be understood as action is intermaterial vibration. An organological inquiry into intermaterial vibration would thus include an inquiry into bodies' oscillatory motion, the simplest form of which can be expressed as harmonic motion. Sujatha offers some examples of simple harmonic motions, including rotor rotation at constant speed, swing motion, orbiting satellites, and tuning fork vibrations. I draw on Sujatha's work to consider a chain of events from acoustic signal to vibration, created under the general condition of air: muscular activity (energy and vibration) activates the laryngeal area, and vocal sound is produced; sound waves in the vocal tract vibrate through the mouth and propagate through the air. Sound reaches listeners directly through the air, or, after being reflected off walls or other surfaces, interacts with the middle and inner ear and excites further vibrations of the basilar membrane and outer hair cells. An electrochemical reaction takes place, which converts the vibration and impact of the hair cells into potentials transmitted to the central nervous system.[81]

This area of knowledge is heavily relied on by traditional (acoustic) instrument makers and, in any time period, by developers of new instruments—

including those making use of electromagnetic forces. For example, piezo ceramic sensors have been developed to measure pressure, acceleration, strain, or force. The piezo—the name comes from the Greek term for "press" or "squeeze"—converts mechanical vibrations to electrical vibrations, or charge.[82] Piezo microphone technology thus works with intermaterial vibration in taking advantage of a disturbance, change, or fluctuation in a relaxed or natural state or axis (of mass, chemicals, electronics, or air).[83] This process allows audiences to detect the material's vibration sonically.[84]

Generations of artists have been interested in energies in various forms, and an organological study of such energies would take into account the study of unwanted vibration—which, within the framework of the figure of sound, is often referred to as "noise"—as well as the study of generating vibration.[85] In the former area, an understanding of vibration's natural occurrence is sought to, for example, build bridges and other structures without inadvertently constructing one that will fail due to naturally occurring vibration. Examples of controlled levels of vibration include drills, engines, and trains traveling on tracks. Indeed, "if the frequency of excitation of a structure coincides with any of its natural frequencies, resonance occurs." Hence, the "failure of major structures like bridges, aeroplane wings and buildings is due to resonance."[86] A 1940 incident that became a recent meme is the YouTube video of the Tacoma Narrows Bridge collapse.[87]

We may recall that "any system which possesses mass and elasticity is capable of vibrating."[88] The study of vibration is divided into two major areas: free and forced vibration.[89] Sujatha explains: "Free vibration is due to forces inherent in the system, while forced vibration is due to externally impressed forces."[90] Resonance occurs when a frequency coincides with a structure's range of excitation. For example, the "typical natural frequency" of a variety of "systems" (the term Sujatha uses for what we might call a "thing") ranges from an offshore oil rig (1 or 2 hertz), a bridge (1–10 hertz), the human trunk (2.5–5 hertz), human vertebra and an SUV's compartment cavity (100 hertz), and a hard disk drive and turbine blade at a high-pressure stage (100–1,000 hertz).[91] Since any body having mass and elasticity has the potential for oscillatory motion, we can appreciate the continuous field through which music is realized.

It is the body's potential for oscillatory motion on which a piece of equipment, a musical instrument such as the piezo microphone, traditional instruments, and vocal and listening organs and bodies and the mass within which they are situated capitalize. And it is this phenomenon that reconfigures our understanding of the boundary objects and related liminal spaces between violin, voice, ear, air, water, or whatever mass transmits the vibrations to which we attend, or that we actively create. It is these spaces between that are broken down under the organology of intermaterial vibration framework.[92]

To be sure, my preliminary suggestions for an organological inquiry into intermaterial vibration are only sketches for where an investigation that goes beyond the figure of sound might ultimately find itself. However, I hope to have offered provocative examples that can suggest some of the ramifications of nudging investigations into music away from the figure of sound. On the surface level, these sketches simply confirm the clear diagnostic forwarded by Brian Kane. "The way to argue against the ideology of sound-in-itself is to demonstrate that sound is always already social."[93] I second Kane and add that the act of identifying a sonic entity, and indeed the ability to do so, is a process that is always already dependent on participation in the "cultural lifeworld."[94]

Kane calls for an investigation that "specif[ies] the relation between forms of sociality and the sounds made."[95] While I also agree fully with this position, what *Sensing Sound* seeks to communicate is that, even in the act of formulating the proposal that the thick event is sound, the figure of sound—a designation assigned by the person interacting with that thick phenomenon—has been set. A person's participation in the "cultural lifeworld" would be facilitated by subscribing to the notion that the thick event of music is best understood as (and limited to) sound. If we add to Kane's diagnostics the recollection of Benjamin Piekut's poignant formulation, "every musical performance is the performance of a relationship," we may begin to sense the ramifications of conceiving music through the figure of sound.[96] The practice of imagining the thick event as knowable sound is performed through naming, singing (or playing), and listening, and it takes place simultaneously with, prior to, and through the "forms of sociality and the sounds made" that Kane urges us to investigate.

The ramifications lie beyond those involved in an inquiry into music per se. The world is, in large part, mediated through a given naturalized acoustic reality.[97] As listeners, we are always positioned in relation to the world via naturalized concepts through which it is formed, and thus, we end up rendering ourselves in these static terms. Moreover, our relations to others are rendered through naturalized modes. Such static terms are proxies for the thick event and stand in the way of inquiry into the complexities of intermaterial relations.

Relationality

I began this book by recalling the question about the sound of a tree falling in a forest. Shifting our attention from whether the falling tree makes a sound if no one is there to hear it to the observation that there will be no sound unless the tree actually falls makes it clear that there must be a physical act that produces sound. Up to this point, this project has posited that the physical act warrants as much attention as the resulting phenomenon; indeed, I have suggested that the event and the practice of sound production occur prior to sound's presence. At the moment when you hear a sound, the forces that initiate it, shape it, and propel it into existence have already carried out their work. We have come to understand that the sound you hear at any given moment therefore cannot be changed: subsequent sounds can be made to differ from previous ones only by adjusting the actions that precede them. Just as the tree must fall to make sound possible, the action of singing must take place for there to be a subsequent vocal sound. From this position, we have drawn the conclusion that what we call singing is therefore not only a matter of sound, but it may result in sound. And I have suggested that, in terms of impact, the sound produced is secondary to the action that produced it.

But if singing is activity alone, and if the sounds produced are secondary, of what does listening consist? Instead of defining singing as the sounds produced by vocal cords and listening as their reception by eardrums, I have proposed that singing and listening are continuously unfolding physical activities and experiences that engage the total human body. We are never privy to sound in the form in which it is transduced through another person's materially specific body. That is, including culturally and historically situated physical senses in deliberations about sound helps inform and shape not only the discussion of our experience of sound but also each specific, embodied, sensory sonic experience. This suggestion is not without precedent. As I have discussed throughout the book, others—including Judith Becker, Steven Connor, Tia DeNora, Steve Goodman, Tomie Hahn, Julian Henriques, Charles Hirschkind, Seth Kim-Cohen, Matt Rahaim, and Michel Serres—have similarly addressed listening's multisensory nature.[98]

When we trace singing and listening back to their grounding in action through an organological approach, we may come to understand that they are conjoined activities. Sound ultimately reverberates throughout the body that hears it; by definition, this is the process that allows a listener to hear a sound. Recall that we are never privy to sound as it is transduced through another person's materially specific body; therefore, sound cannot be reduced to an objective index or object. Moreover, tracing singing and listening to action shows us that listening is not passive in its relationship to singing: the listener does not merely receive what he or she hears. As a result, listening, like singing, should be defined as the transmission and transduction of sound.

We may therefore think about propagation, transmission, and transduction—specifically the transduction of aural vibration, especially in the 20 hertz to 20 kilohertz range—as the common denominator of material mass, muscle activity, sound, and hearing: the elements that constitute singing.[99] If singing and listening are the actions that give rise to sound—in the vibration that surges through the singer, and in the material that envelops the singer and listener—does this sound, this vibration, have a beginning or end? It does not. The vibration is expressed as transmission or as transduction, and depending on how we define a node within that continuous field, we may define its beginning and end. But the vibration or energy in itself does not imply or express a bounded object with a beginning or ending point.

The following formula may help us to understand these interlocking relationships and summarize the argument I have made thus far in the book:

If [s] and [l] are [v] and [v] is [r];
And, if [b] is [v];
Then, also [b] is [r].

[s] = singing; [l] = listening; [v] = vibration across bodies, causing change; [b] = being; [r] = relational

Singing and listening are particular expressions of the processes of vibration. What we understand as sound ultimately reverberates throughout the material body that produces and senses it; it is precisely because sound—undulating energy—is transduced through the listener's body that it is sensed. On the one hand, when we produce music we ourselves are affected by the process. On the other hand, by projecting music out into the air, we have an impact on the world around us. We do not engage with music at a distance but, by definition, we do so by entering into a relationship that changes us. The most extreme definition of music possible, then, is vibrational energy—and, at times, transformation through that vibrational energy, which is an always already unfolding relational process.

Music arises in the confluence between the materiality we offer up and the vibrational force that is put forth into the world. As a consquence, (1) to participate in music is to offer oneself up to that music; (2) to put music forth into the world is to have an impact on another; and, therefore, (3) it is as propagators and transductional nodes of that thick event of music—the full vibrational range, including sub- and ultrafrequencies—that we participate in and are privy to music.[100] What connects singing, listening, and sound, then, is vibration. Indeed, what connects the physical, full-body activities and experiences that take place during both singing and listening is the transmission and transduction of vibration. That is, if music is not something external and objective but is transmitted from one material node to another, music indeed puts us into an intrinsic dynamic, material relationship to both the so-called external world and each other. Musical discourse then shifts from the realm of the symbolic to that of the relational.

From the Relational to the Thick Event

Throughout this book, I have continually returned to the relational aspects of sound, music, singing, and listening. Typically, when people discuss a particular facet of what I have called the thick event, there is an assumption that it constitutes the entirety of that event—for example, the sound of a falling tree, which is actually only one of the many aspects of the full experience of a tree falling in the forest. If we make this assumption, we may think that understanding the nuances of a particular facet of an experience can provide an understanding of the thick event. However, if we explore each isolated phenomenon as a way of grasping one aspect of the thick event, we realize that no single aspect is the event. The process of putting these aspects, traditionally understood in isolation, into relationship with each other foregrounds the absences, lacunae, or shadows that suggest a more complex event (that is, the thick event). And this, in turn, emphasizes the isolation of each aspect and the fact that one aspect cannot be dealt with alone, if we wish to engage the full event.

In this way, relationality erases itself. By first showing the multiple "slices" involved in the whole, and how they interact as a whole, relationality leads us back to dealing with the whole as an experience, rather than as an isolated idea. As we set up a process that prompts thinking relationally through aspects of sound, music, singing, and listening, this linking presumes isolated aspects that are connected. In other words, once the thick event is recognized experientially, then a return to naming conventions, to using pieces to represent the whole, diverts our attention from the next step: dealing with the rediscovered thick event as intermaterial vibrations within which we ourselves are situated, and to which we ourselves contribute.

In chapter 4, I considered how a thick event could be reduced and naturalized. For example, I discussed the breathing exercise commonly given to classical singers that instructs them to inhale as if they were smelling a rose. While I deduced that this exercise was devised from a careful analysis of correct inhalation technique and that its naming served to enable future repetition and pedagogy, the exercise favors certain aspects of the thick inhalation event and isolates others from its complexity. Moreover, this process of selection and its naming often ends up replacing the full action on which it was based. That is, the naming and the directions designed to accomplish the exercise seem to suggest that the name itself—not the full event on which it was based—is the goal.

In the same way that the direction to "inhale as though you are smelling a rose" is an approximation of a full inhalation, the term and concept of vibration is a placeholder for something that is much fuller and more complex than any name and concept can possibly contain. As I have discussed throughout the book, and in this chapter in particular, interdisciplinary material investigation of some musical parameters led me to this concept and term. However, in the same way that relying on the "smelling a rose" exercise can lead to the complete inhalation from which the exercise was first derived, describing the operatic works of Juliana Snapper and Meredith Monk as intermaterial vibrational unfoldings fleshes out naturalized concepts such as sound but still reduces the thick event. In other words, to reduce opera or other musical events to verifiable parameters hardly captures the thick event. As this book has sought to illustrate, setting performances in different material and spatial configurations, denaturalizing musical parameters through performance, or actualizing explications rather than paraphrasing them into confirmable parameters may all serve to retain the thick event.

Unexpectedly, then (at least for me), an investigation of twenty-first-century opera does not necessarily bring us back to a study of or about music. Instead, it has taken me to the location of the artists and musicians themselves. The word *poetry* names the very essence of art (namely, *poiêsis* or "bringing into being"),[101] hence Heidegger claims that "*All art . . . is . . . essentially poetry.*"[102] This suggests not only that interacting with music and seeking to engage it is art, but also that engaging the music itself—and the voices, artists, and musicians themselves—may bring us closest to the knowledge about the music.

As soon as we attempt to name and explicate the thick event—that which is brought into being—we position ourselves outside it and, by definition, approach it from a single point of view or engage it through one dominant sense—in which case we are already defining the thick event as sound (for example) and thus asking questions about just one of its aspects. As I have discussed throughout the book, through this process the thick event is carefully parsed in the interest of arriving at measured and controlled verifiable knowledge. In contrast, the current that swept me up while heeding the call of the artists and the music studied here led me to want to contribute to bringing into being. We may then revise our formula:

If [r] serves to show the connection between
    naturalized [s], [l], and [v],
when we pay attention to the [te],
[r] is already contained and subsumed within the [te].

[s] = singing; [l] = listening; [v] = vibration across bodies, causing change;
[b] = being; [r] = relational; [te] = thick event

Singing and listening are aspects of a thick event and are distinguished from it only through naming. In the same way, relationality through intermaterial vibration is only a placeholder for the thick event. As Luigi Russolo puts it, "to enrich means to add, not to substitute or to abolish."[103]

Everything and Nothing, and Not Even That

The sonorous . . . outweighs form. It does not dissolve it, but rather enlarges it; it gives it an amplitude, a density, and a vibration or undulation whose outline never does anything but approach. The visual persists until its disappearance; the sonorous appears and fades away into its permanence. —JEAN-LUC NANCY, *Listening*

In reading contemporary vocal works as materially situated and realized, *Sensing Sound* has offered analytical perspectives intended to develop a greater understanding of the nuanced range of human encounters with music and has proposed that certain aspects of human relations reveal themselves poignantly in song. Considering Jean-Luc Nancy's formulation, "The sonorous . . . outweighs form,"[104] I have demonstrated that, in encounters through and with music, we are physically touched and we tangibly touch others. Whether we are performing, listening, or engaging in scholarship, what is at stake in music is nothing less than the fundamental human experience of touching and being touched. It is precisely because music is the site of such transactions that, even when disciplines beyond music study it closely, music yields unique knowledge. While conceptualizing music through the figure-of-sound paradigm delineates the limits of musical phenomena and the insights that they can produce, conceptualizing music as material transduction and physical touch opens the inquiry to transdisciplinary relevance and consequence.

However, if I take on my own proposition that the designations *sounds, musical parameters*, and *musical entities* are intermaterial vibration—or, better yet, thick events—in understanding that they only come into relief through identification with a priori parameters, there is still a crucial area on which I have yet to turn a critical organological lens. Returning to the falling tree, I have already identified what the classic question implies: it severs certain parts of the vibration, inserting them into the identifier *sound*. I also argue in that regard that, if you were there, you might be more concerned with the possibility that the tree would fall on you than with the sound. Focusing more narrowly on this question, we may continue to apply the material vibrational organological lens. In my assumption that something—a human—can be pinned underneath the tree lies the presumption of a knowable entity, that of the human being. This conceptualization presumes identification between material energy and human form. Such an assumption implicitly asks: Where does the human being begin and end? Where is the perimeter of the potentially injured entity? What holds it together as a unified entity? What is it that could be injured? Is the skin cell that falls off immediately prior to the tree's descent injured? Is the hair injured? Are the atoms that make up a cracked bone injured? If we believe that these aspects do not make up an injured entity, why is this? And if we do not doubt, and we sense that a human being could indeed be injured, why is that?

In understanding vibrations as both sound and aspects of the thick event, the designation *human entity* is thrown into relief following a preconceived concept of human form. There is no essential human form besides an a priori idea that holds that form together.[105] In other words, without this engaged idea, we would not hold the reference according to which we see a human form pinned underneath a tree. As the question of the falling tree's possible sound is both limited and directed by deep assumptions regarding the ontology of sound and music, so is the idea that it is a human who would be pinned under the tree. As I hinted at above, if we were to apply a different scale to the scene, we would not form an understanding that what we witnessed was a human body under the tree. On a scale with higher resolution, we would understand only that molecules were shifting. On a broader scale, we would understand only the relation between shifting landmasses.

Therefore, mapping the dynamics arising between nodes that can be understood to form relationships is merely mapping the relationships between ideas fixed to a priori ideals and confined segments of the continuous field of vibration and energy. If we question identification, releasing its functional power, there is no one stable and knowable music that has the capacity to both restore and destroy. As we move toward the end of this book, what I suggest is this: if there were no identifying a priori orientations—not only for the building blocks we understand as constituting music, but also for the entity we understand as experiencing and making meaning through that music—we would be unable even to identify that entity as a human being. And in that case, there would be no entity that could be injured by music. Without identification, not only does the illusion of a single stable and knowable music evaporate, but also there is no human being to be restored or destroyed by it.

In becoming aware of the power of identification, we may come to realize that what we have identified as a human body could as easily be identified as flesh, bone, or skin cells, thousands of which detach themselves from the so-called human body every minute. And when we stand on the ground, if we release the anthropocentric framework that divides our perception into identifications of human flesh and dirt, we could understand ground, feet, nails, ankles, knees, and so on as a continuous field of atoms.

If we can occupy that porous and unbounded place, unattached to and thus released from identification and its ramifications; if we can participate in a zone of experience where a priori identifications dissolve;[106] then we will participate in the thick event as everything, and as nothing—within a mode in which not even that is true.[107]

NOTES

Introduction

1. I have not been able to locate a single origin for ponderings about trees and perception. Without mentioning the falling tree, George Berkeley does call upon the tree in *Principles of Human Knowledge* (1710), section 45: "The objects of sense exist only when they are perceived: the trees therefore are in the garden, or the chairs in the parlour, no longer than while there is somebody by to perceive them" (George Berkeley, *Principles of Human Knowledge and Three Dialogues between Hylas and Philonous*, ed. Roger Woolhouse [London: Penguin, 1988], 68). An 1883 posing of the question, "If a tree were to fall on an island where there were no human beings would there be any sound?" can be found (with no author given) in a magazine ("Editor's Table," *Chautauquan*, June 1883, 543–44). The reply to the question is: "No. Sound is the sensation excited in the ear when the air or other medium is set in motion." Rewording the question, in 1884 *Scientific American* discarded the philosophical aspect and picked up on the scientific and technical strain of this question: "If a tree were to fall on an uninhabited island, would there be any sound?" And the anonymous author gave a more technical answer: "Sound is vibration, transmitted to our senses through the mechanism of the ear, and recognized as sound only at our nerve centers. The falling of the tree or any other disturbance will produce vibration of the air. If there be no ears to hear, there will be no sound" ("Notes & Queries," *Scientific American* 50, no. 14 [1884], 218). The current phrasing might have originated around a century ago, with "When a tree falls in a lonely forest, and no animal is near by to hear it, does it make a sound? Why?" (Charles Riborg Mann and George Ransom Twiss, *Physics* [Chicago; New York: Scott, Foresman, and Company, (1905) 1910], 235).

2. Clifford Geertz borrows the term *thick description* from Gilbert Ryle to specifically address issues related to ethnography. *The Interpretation of Cultures* (New York: Basic, 1973), 6.

3. The discussion and statements herein concern Western concert music in general and opera within Western culture in particular. Moreover, I would like to note that I recognize that these traditions are more complex and diverse than I can always account for.

4. The term *figure of sound*, of course, conceptually and sonically resembles the term

figure of speech. For issues specifically connected to conceptualization of sound in new media, see Norie Neumark, Ross Gibson, and Theo Van Leeuwen, *Voice: Vocal Aesthetics in Digital Arts and Media* (Cambridge, MA: MIT Press, 2010).

5. This points toward a way in which this methodological and theoretical implication can assist work on difference. I do not address this directly herein, but I have started to hint at it elsewhere (Nina Sun Eidsheim, "Voice as Action: Towards a Model for Analyzing the Dynamic Construction of Racialized Voice," *Current Musicology* 93, no. 1 [2012]: 7–31), and I will treat it thoroughly in a monograph in process (Nina Sun Eidsheim, "Measuring Race: Listening to Vocal Timbre and Vocality in African-American Popular Music," unpublished manuscript).

6. This approach and perspective is much indebted to Christopher Small, *Music, Society, Education* (Hanover, NH: University Press of New England, 1996).

7. A number of individual authors and texts could be mentioned in this context, and they are referred to throughout this book. However, I will acknowledge here the ways in which sound studies and certain areas of science and technology studies have reinvigorated music scholarship. The arrival of sound studies has finally been marked by institutional recognition, including in organizations (the European Sound Studies Association), edited volumes (Trevor Pinch and Karin Bijsterveld, eds., *The Oxford Handbook of Sound Studies* [New York: Oxford University Press, 2012]; Jonathan Sterne, ed., *The Sound Studies Reader* [New York: Routledge, 2012]), and journals (*Interference: A Journal of Audioculture*; the *Journal of Sonic Studies*; *SoundEffects: An Interdisciplinary Journal of Sound and Sound Experience*; *Sound Studies: An Interdisciplinary Journal*). However, as I hope will become clear, while this project would unequivocally not have been possible without the perspectives offered by work within sound studies, *Sensing Sound* offers a challenge to how sound studies defines its object.

I have also greatly benefited from, and continue to build on, work in the emerging area of voice studies. Recent key works include Shane Butler, *The Ancient Phonograph* (New York: Zone Books, 2015); Olivia Ashley Bloechl, *Native American Song at the Frontiers of Early Modern Music* (Cambridge: Cambridge University Press, 2008); Adriana Cavarero, *For More Than One Voice: Toward a Philosophy of Vocal Expression* (Stanford, CA: Stanford University Press, 2005); Steven Connor, *Beyond Words: Sobs, Hums, Stutters and other Vocalizations* (London: Reaktion Books Ltd., 2014), and *Dumbstruck: A Cultural History of Ventriloquism* (Oxford: Oxford University Press, 2000); James Q. Davies, *Romantic Anatomies of Performance* (Berkeley: University of California Press, 2014); Frances Dyson, *Sounding New Media: Immersion and Embodiment in the Arts and Culture* (Berkeley: University of California Press, 2009); Nina Sun Eidsheim and Katherine Meizel, eds., *The Oxford Handbook of Voice Studies* (New York: Oxford University Press, forthcoming); Martha Feldman, *The Castrato: Reflections on Natures and Kinds* (Berkeley: University of California Press, 2014); Martha Feldman and Bonnie Gordon, eds., *The Courtesan's Arts: Cross-Cultural Perspectives* (New York: Oxford University Press, 2006); Bonnie Gordon, *Monteverdi's Unruly Women: The Power of Song in Early Modern Italy* (Cambridge: Cambridge University Press, 2004); Krzystof Izdebski, *Emotions in the Human Voice* (San Diego, CA: Plural, 2008); Brian Kane, *Sound Unseen: Acousmatic Sound in Theory and Practice* (New York: Oxford University Press, 2014); Jody Kreiman and Diana Sidtis, *Foundations of Voice*

Studies: An Interdisciplinary Approach to Voice Production and Perception (Malden, MA: Wiley-Blackwell, 2011); Brandon LaBelle, *Lexicon of the Mouth: Poetics and Politics of Voice and the Oral Imaginary* (New York: Bloomsbury, 2014); Katherine Meizel, *Idolized: Music, Media, and Identity in* American Idol (Bloomington, IN: Indiana University Press, 2011); Ana María Ochoa Gautier, *Aurality: Knowledge and Listening in Nineteenth-Century Colombia* (Durham, NC: Duke University Press, 2014); Matthew Rahaim, *Musicking Bodies: Gesture and Voice in Hindustani Music* (Middletown, CT: Wesleyan University Press, 2012); Jacob Smith, *Vocal Tracks: Performance and Sound Media* (Berkeley: University of California Press, 2008); Gary Tomlinson, *The Singing of the New World: Indigenous Voice in the Era of European Contact* (Cambridge: Cambridge University Press, 2007); Amanda J. Weidman, *Singing the Classical, Voicing the Modern: The Postcolonial Politics of Music in South India* (Durham, NC: Duke University Press, 2006); the special issues of the journals *Qui Parle* (Simon Porzak, ed., "Liner Notes: The Margins of Song" [special issue, *Qui Parle* 21, no. 1 (2012)]) and *Postmodern Culture* (Nina Sun Eidsheim and Annette Schlichter, eds., special issue on voice and materiality, forthcoming).

8. Suzanne Cusick discusses how the "manipulations of the acoustic disrupted prisoners' use of hearing and vocalization both to locate themselves in intelligible worlds and to create relationships with those worlds," observing that "the destruction of prisoners' subjectivities partly depends on the acoustically and philosophically salient fact that manipulations of the acoustical environment always produce the somatic effect of sympathetic vibrations." Finally, she argues that in being subjected to "music's acoustical energy," the prisoners are left without "the capacity to control the outside relationality that is the foundation of subjectivity" (Suzanne G. Cusick, "Acoustemology of Detension in the 'Global War on Terror,'" in *Music, Sound and Space: Transformations of Public and Private Experience*, ed. Georgina Born [Cambridge: Cambridge University Press, 2013], 276). In other words, we define ourselves through acoustic practice, and our selves are eradicated when the ability to control it is taken away. In the introduction to the same volume, Georgina Born discusses "music, sound and space" as "three kinds of irreducible multiplicity at work in the musical and sonic experience" (introduction to *Music, Sound and Space: Transformations of Public and Private Experience*, ed. Georgina Born [Cambridge: Cambridge University Press, 2013], 19).

9. Geertz, *The Interpretation of Cultures*, 10.

10. For a discussion about measuring sound as a feature that marked modernity, see Joseph Auner, "Weighing, Measuring, Embalming Tonality: How We Became Phonometrographers," in *Tonality 1900–1950: Concept and Practice*, ed. Feliz Wörner, Ullrich Scheideler, and Philip Rupprecht (Stuttgart, Germany: Franz Steiner Verlag, 24–46.

11. This is further complicated in contexts where music literacy is understood as textual literacy.

12. Billy Collins, "TED Radio Hour," National Public Radio, June 1, 2012, accessed April 17, 2015, http://www.npr.org/templates/transcript/transcript.php?storyId =153699514.

13. Collins's sentiment is close to the position outlined by Susan Sontag in her now-classic essay "Against Interpretation." In it she summarized her call for independence from the established obligations of criticism and scholarship in her infamous exhorta-

tion to replace hermeneutics with "an erotics of art" (*Against Interpretation and Other Essays* [New York: Farrar, Straus, and Giroux, (1961) 1966], 14).

14. The issue of aurality in relation to music is further complicated in discussions of the musical work. For many, locating the musical work can be based purely on (visual) textual literacy and competencies. This complex discussion lies beyond the scope of this study. In using the term *paradigm*, I follow Thomas J. Csordas's usage: "By paradigm I mean simply a consistent methodological perspective that encourages re-analyses of existing data and suggests new questions for empirical research" ("Embodiment as a Paradigm for Anthropology," *Ethos* 18, no. 1 [1990]: 5).

15. See Christopher Small, *Musicking: The Meanings of Performing and Listening* (Middletown, CT: Wesleyan University Press, 1998). Small has not been alone in drawing our attention to the interconnectedness between so-called music and so-called nonmusical events. However, his corralling of a much broader range of activities under the term *musicking* has been very useful. Staying within music studies, other insightful approaches include those of ethnomusicologists, such as Steven Feld, *Sound and Sentiment: Birds, Weeping, Poetics, and Song in Kaluli Expression*, 2nd ed. (Philadelphia: University of Pennsylvania Press, 1990); Anthony Seeger, *Why Suyá Sing: A Musical Anthropology of an Amazonian People* (Cambridge: Cambridge University Press, 1987). More recently, Rahaim specifically employs the term *musicking* to capture the connection between gesture and voice (*Musicking Bodies*).

16. Peter Szendy, *Listen: A History of Our Ears*, preceded by "Ascoltando" by Jean-Luc Nancy; translated by Charlotte Mandell (New York: Fordham University Press, 2008), 7.

17. Eidsheim, "Voice as Action."

18. Marion A. Guck, "Who Counts?" (paper presented at the joint meeting of the American Musicological Society, Society for Ethnomusicology, and Society for Music Theory, New Orleans, LA, November 3, 2012). While Georgina Born's call for a "relational musicology" concerns "music scholarship" that is "no longer imprinted with the disciplinary assumptions, boundary and divisions inherited from the last century" ("For a Relational Musicology: Music and Interdisciplinarity, Beyond the Practice Turn," *Journal of the Royal Music Association* 135, no. 2 [2010]: 205, 242), and this book advocates a material relational account of music and the people involved in its production and reception, the two perspectives are related. It is in disciplinary assumptions that objects of analysis are defined and separated into isolated areas of discourse.

19. Theodor Adorno, *Essays on Music*, ed. Richard Leppert and trans. Susan H. Gillespie (Berkeley: University of California Press, 2002), 452.

20. See Larry Laskowski, *Heinrich Schenker, an Annotated Index to His Analyses of Musical Works* (New York: Pendragon Press, 1978), for a complete list of works discussed by Schenker.

21. Jonathan Sterne, *The Audible Past: Cultural Origins of Sound Reproduction* (Durham, NC: Duke University Press, 2003), 103.

22. Jon Cruz, *Culture on the Margins: The Black Spiritual and the Rise of American Cultural Interpretation* (Princeton, NJ: Princeton University Press, 1999), 38, 7.

23. Mark M. Smith, *Listening to Nineteenth-Century America* (Chapel Hill: University of North Carolina Press, 2001), 7.

24. In this book, by the term *a priori* I refer to a priori justification relying on non-experiential sources of evidence and a priori knowledge as based on a priori justification. For further explication, see Bruce Russell, "*A Priori* Justification and Knowledge," Stanford Encyclopedia of Philosophy Archive, summer 2014, accessed November 1, 2014, http://plato.stanford.edu/archives/sum2014/entries/apriori/.

25. See Bryan Gick and Donald Derrick, "Aero-Tactile Integration in Speech Perception," *Nature* 462, no. 7272 (2009): 502–4.

26. Matthew Rahaim discusses the intertwining of vocality, ethics, and morality in *Musicking Bodies*, 126–32, and in his forthcoming book, tentatively titled "Voice Cultures: Varieties of Ethical Power in Hindustani Music" (unpublished manuscript).

27. However, others have written historical studies about closely related topics, and these brilliant intellectual undertakings have made this particular project possible. They include Steven Connor, *Dumbstruck*; Veit Erlmann, *Reason and Resonance: A History of Modern Aurality* (New York: Zone, 2010); Douglas Kahn, *Noise, Water, Meat: A History of Sound in the Arts* (Cambridge, MA: MIT Press, 1999); Seth Kim-Cohen, *In the Blink of an Ear: Towards a Non-Cochlear Sonic Art* (New York: Continuum, 2009); Mara Mills, "Deaf Jam: From Inscription to Reproduction to Information," *Social Text* 28, no. 102 (2010): 35–58; Peter Price, *Resonance: Philosophy for Sonic Art* (New York: Atropos, 2011); Hillel Schwartz, *Making Noise: From Babel to the Big Bang and Beyond* (New York: Zone, 2011); M. Smith, *Listening to Nineteenth-Century America*; Jonathan Sterne, *The Audible Past* and *MP3: The Meaning of a Format* (Durham, NC: Duke University Press, 2012); Timothy D. Taylor, *The Sound of Capitalism* (Chicago: University of Chicago Press, 2012); Emily Ann Thompson, *The Soundscape of Modernity: Architectural Acoustics and the Culture of Listening in America, 1900–1933* (Cambridge, MA: MIT Press, 2002).

28. This point is discussed in Kahn, *Noise, Water, Meat*, 200. My thanks to Brian Kane, who reminded me of this.

29. José Esteban Muñoz, *Disidentifications: Queers of Color and the Performance of Politics* (Minneapolis: University of Minnesota Press, 1999), 26.

30. Ibid., 28.

31. Importantly, while my deconstruction of naturalized musical parameters and concepts arrives at the practice of vibration, I never offer vibration as essence. I rely on vibration only as a placeholder for that which cannot be named: each unrepeatable relationship and idiosyncratic experience. It is precisely naming that would fix the experience, which would then be limited by, and would espouse the assumptions of, the given paradigm from which its name arose.

32. Some readers may expect references to and explicit engagement with music psychology and cognitive science scholarship. In anticipation of such responses, I want to address this expectation. In the research stage of this project, I did explore the literature on music and affect within music psychology and cognitive science, and this literature indeed forms the project's backbone. However, after immersing myself in that realm of thought for over two intense years, I learned that my argument about the embodied and

material dimensions of music may in fact be made in an equally powerful way without directly referring to neuroscientific research or deferring to knowledge within the neurosciences. Thus, I take a different path (to the same end).

33. Indeed, it was my work on perception and production of race and vocal timbre that triggered the questions asked in this project. According to which principles do we delineate what we think of as a given sound, or the quality of a given sound? And according to which values and principles do we derive meaning and affect from such delineation?

34. This scholarship includes Carolyn Abbate, *In Search of Opera* (Princeton, NJ: Princeton University Press, 2001), "Music—Drastic or Gnostic?," *Critical Inquiry* 30, no. 3 (2004): 505–36, and *Unsung Voices: Opera and Musical Narrative in the Nineteenth Century* (Princeton, NJ: Princeton University Press, 1991); Robert Beahrs, "Post-Soviet Tuvan Throat-Singing (*Xöömei*) and the Circulation of Nomadic Sensibility," PhD diss., University of California, Berkeley, 2014; Judith O. Becker, *Deep Listeners: Music, Emotion, and Trancing* (Bloomington: Indiana University Press, 2004); Katherine Bergeron, *Voice Lessons: French Mélodie in the Belle Epoque* (Oxford: Oxford University Press, 2010); Hannah Bosma, "The Electronic Cry: Voice and Gender in Electroacoustic Music," PhD diss., University of Amsterdam, 2013; Zeynep Bulut, "La Voix-Peau: Understanding the Physical, Phenomenal and Imaginary Limits of the Human Voice through Contemporary Music," PhD diss., University of California, San Diego, 2011; Suzanne G. Cusick, "On Musical Performances of Gender and Sex," in *Audible Traces: Gender, Identity and Music*," ed. Elaine Barkin and Lydia Hamessley (Los Angeles: Carciofoli Verlagshaus, 1999), 25–48; Mladen Dolar, *A Voice and Nothing More* (Cambridge, MA: MIT Press, 2006); Leslie C. Dunn and Nancy A. Jones, *Embodied Voices: Representing Female Vocality in Western Culture* (Cambridge: Cambridge University Press, 1994); Nina Sun Eidsheim, "Sensing Voice: Materiality and Presence in Singing and Listening," *Senses & Society* 6, no. 2 (2011): 133–55, and "Voice as a Technology of Selfhood: Towards an Analysis of Racialized Timbre and Vocal Performance," PhD diss., University of California, San Diego, 2008; Eidsheim and Schlichter, eds., *Postmodern Culture*, special issue on voice and materiality; Gordon, *Monteverdi's Unruly Women*; Elisabeth Le Guin, *Boccherini's Body: An Essay in Carnal Musicology* (Berkeley: University of California Press, 2006); Susan McClary, *Feminine Endings: Music, Gender, and Sexuality* (Minneapolis: University of Minnesota Press, 1991), and *Modal Subjectivities: Self-Fashioning in the Italian Madrigal* (Berkeley: University of California Press, 2004); Katherine Meizel, "A Powerful Voice: Investigating Vocality and Identity," *Voice and Speech Review* 7, no. 1 (2011): 267–74; Porzak, ed., "Liner Notes"; Szendy, *Listen*.

35. The following works include some of the exceptions: Genevieve Calame-Griaule, "Voice and the Dogon World," in *Notebooks in Cultural Analysis*, ed. Norman F. Cantor and Nathalia King (Durham, NC: Duke University Press, 1986), 15–60; Steven Connor, *Dumbstruck* and "Edison's Teeth: Touching Hearing," in *Hearing Cultures: Essays on Sound, Listening and Modernity*, ed. Veit Erlman (Oxford: Berg, 2004), 153–72; Emma Dillon, *The Sense of Sound: Musical Meaning in France, 1260–1330* (New York: Oxford University Press, 2012); Dyson, *Sounding New Media*; Feld, *Sound and Sentiment*; Stefan Helmreich,

"An Anthropologist Underwater: Immersive Soundscapes, Submarine Cyborgs, and Transductive Ethnography," *American Ethnologist* 34, no. 4 (2007): 621–41, and "Underwater Music: Tuning Composition to the Sounds of Science," in *The Oxford Handbook of Sound Studies*, ed. Trevor Pinch and Karin Bijsterveld (New York: Oxford University Press, 2012), 151–75; Julian Henriques, "The Vibrations of Affect and Their Propagation on a Night out on Kingston's Dancehall Scene," *Body and Society* 16, no. 1 (2010): 57–89; Don Ihde, *Listening and Voice: A Phenomenology of Sound* (Athens: Ohio University Press, 1976); Kahn, *Noise, Water, Meat*; Brandon LaBelle, *Acoustic Territories: Sound Culture and Everyday Life* (New York: Continuum, 2010) and *Background Noise: Perspectives on Sound Art* (New York: Continuum International, 2006); Brian Massumi, *Parables for the Virtual: Movement, Affect, Sensation* (Durham, NC: Duke University Press, 2002); Ingrid Monson, "Hearing, Seeing, and Perceptual Agency," *Critical Inquiry* 34, no. 5 (2008): S36–58; Jean-Luc Nancy, *Listening*, trans. Charlotte Mandell (New York: Fordham University Press, 2007); Tara Rodgers, "Synthesizing Sound: Metaphor in Audio-Technical Discourse and Synthesis History," Ph.D. diss., McGill University, 2010, and "Toward a Feminist Epistemology of Sound: Refiguring Waves in Audio-Technical Discourse," in *Philosophy after Irigaray*, ed. Mary Rawlinson, Danae Mcleod, and Sara McNamara (Albany: State University of New York Press, forthcoming); Matt Sakakeeny, "'Under the Bridge': An Orientation to Soundscapes in New Orleans," *Ethnomusicology* 54, no. 1 (2010): 1–27; Michel Serres, *The Five Senses: A Philosophy of Mingled Bodies*, trans. Margaret Sankey and Peter Cowley (New York: Continuum, 2009); Seeger, *Why Suyá Sing*.

36. Here I think about pitch-, rhythm-, harmony-, and form-oriented analysis that grew out of a quest to understand a particular repertoire from which the analytical values and perspectives flowed. For instance, the value given to pitch coherence and harmonic logic and complexity grew out of a limited repertoire that exhibited those specific qualities. However, in my view, it is limiting—even futile—to apply those values and instruments to music outside their orbit. Additionally, the values account for only a limited aspect of even the repertoire that they fit most perfectly. Richard Middleton refers to this as "the musicological problem" (*Studying Popular Music* [Buckingham, UK: Open University Press, 1990], 103). See also David Brackett, *Interpreting Popular Music* (Cambridge: Cambridge University Press, 1995), 19; Susan McClary and Robert Walser, "Start Making Sense: Musicology Wrestles with Rock," in *On Record: Rock, Pop, and the Written Word*, ed. Simon Frith and Andrew Goodwin (New York: Pantheon, 1990), 237–49.

37. Abbate, "Music—Drastic or Gnostic?," 514.

38. In a response to Abbate, Emma Dillon (*The Sense of Sound*) points out that a number of disciplines have been in pursuit of the drastic for a long time. Both ethnomusicology and scholarship on performance, including performance studies, are mindful of the "live" aspects of music that "unfold[s] in time," for which Abbate called ("Music—Drastic or Gnostic?," 505).

39. Feldman, *The Castrato*; Emily Dolan, "Perspectives on Critical Organology," *Newsletter of the American Musical Instrument Society* 43, no. 1 (2014): 14–16.

40. Mills, "Deaf Jam."

41. Erlmann, *Reason and Resonance*.

42. Quoted in Auner, "Weighing, Measuring, Embalming Tonality," 25, 26.

43. Alain Corbin, *Village Bells: Sound and Meaning in the 19th-century French Countryside* (New York: Columbia University Press, 1998).

44. Including Karen Michelle Barad, *Meeting the Universe Halfway: Quantum Physics and the Entanglement of Matter and Meaning* (Durham, NC: Duke University Press, 2007); Estelle Barrett and Barbara Bolt, *Carnal Knowledge: Towards a "New Materialism" through the Arts* (London; New York: I.B. Tauris, 2013); Gaile Sloan Cannella, Michelle Salazar Pérez, and Penny A. Pasque, *Critical Qualitative Inquiry: Foundations and Futures* (Walnut Creek, CA: Left Coast Press, 2015); Diana H. Coole and Samantha Frost, *New Materialisms: Ontology, Agency, and Politics* (Durham, NC: Duke University Press, 2010); Jane Bennett, *Vibrant Matter: A Political Ecology of Things* (Durham, NC: Duke University Press, 2010).

45. Le Guin, *Boccherini's Body*, 14, see also 15–37; Davies, *Romantic Anatomies of Performance*, 2; Peter Lunenfeld, "The Maker's Discourse," paper presented at the "Critical Mass: The Legacy of Hollis Frampton" conference, University of Chicago, IL, February 6, 2010. Thanks to Michael D'Errico for pointing me to Lunenfeld's work.

46. Beyond the points I make in *Sensing Sound*, to me the main takeaway lesson from learning about the voice in the vocal studio is its infinite malleability and adaptability. First, the voice has the ability to maintain its personal character (that which allows us to recognize a particular person) while taking on a social or cultural timbral category. Second, because the voice is so malleable, any social or cultural category perceived by the entity that trains the voice can be trained into the voice. This vocal training entity can be a voice teacher, but it can also consist of self-adjustment in search of social recognition. See Nina Sun Eidsheim, "Race and Aesthetics of Vocal Timbre," in *Rethinking Difference in Music Scholarship*, ed. Olivia Bloechl, Melanie Lowe, and Jeffrey Kallberg (Cambridge: Cambridge University Press, 2015), 338–65. This is also the topic of my "Measuring Race."

47. I am grateful to the practice- and experimentally oriented environment, the PhD program called Critical Studies and Experiential Practices (in the Department of Music at the University of California, San Diego) that constituted my first encounter with academia outside of music conservatory setting. There, I was fortunate to be mentored by intellectual powerhouses—who were also deeply committed to artistic and musical practices—including Anthony Davis, Adriene Jenik, George Lewis, George Lipsitz, Jann Pasler, and Miller Puckette.

48. Susan McClary, "Constructions of Subjectivity in Schubert's Music," in *Queering the Pitch: The New Gay and Lesbian Musicology*, ed. Philip Brett, Elizabeth Wood, and Gary Thomas (New York: Routledge, 1994), 211.

49. Small, *Musicking*, 2.

50. Considering music from the point of view of energy occurs throughout history and across cultures. The eclectic and vast bibliography is beyond the scope of this book.

51. The anthropologist Stefan Helmreich (who employs the concept of immersion to think about sound underwater and descent into a "cultural medium"/area of research ["An Anthropologist Underwater," 630]) and the sociologist Adrien MacKenzie (who applies the term *transduction* to metaphorically address technology's effect on bodies

and machines) have also used the concept of vibration, broadly, as a way of bringing different perspectives and disciplines into relationship ("Underwater Music"; Adrian Mackenzie, *Transductions: Bodies and Machines at Speed* [London: Continuum, 2002]).

52. Jane Bennett, *Vibrant Matter: A Political Ecology of Things* (Durham, NC: Duke University Press, 2010), 99. The example Bennett uses is *Great Treatise on Supreme Sound*, which she describes as a fourteenth-century handbook for musicians and which is not widely available.

53. Bennett, *Vibrant Matter*, dust jacket.

54. Rebecca Lippman, "Musicology in the Flesh: a Sensual Inquiry Into Music" seminar, March 12, 2013, University of California, Los Angeles.

55. See, for example, Albert Einstein, "Does the Inertia of a Body Depend upon Its Energy Content?," in *The Collected Papers of Albert Einstein*, vol. 2, *The Swiss Years: Writings, 1900–1909*, trans. Anna Beck (Princeton, NJ: Princeton University Press, 1989), 174. Thanks to William Waters for directing me to this reference.

56. I recognize that vibration is not the only way to get beyond the discourse of fidelity.

57. This becomes particularly problematic in voice when the metaphorical use of voice for subjectivity and interiority is laid on top of this linking of the sound of the voice and signified. I have written about this problem in terms of race. The issue of fidelity arises here in relation to what is expected, as in the voice sounds like a female voice, so the vocalizer is female, or looks black, should sound black. This logic was used in a court case. In 1999, the Kentucky Supreme Court ruled that a conviction was appropriately based solely on a police officer's identification of a suspect whose voice the officer heard on an audio transmission. The officer identified the suspect as a black male and testified that during his thirteen years as a policeman he had had several conversations with black men and therefore was able to identify the voice of a black male. In his ruling, the judge deduced that no one would find it inappropriate for an officer to identify the voice of a woman, and hence, "we perceive no reason why a witness could likewise identify a voice as being that of a particular race or nationality, so long as the witness is personally familiar with the general characteristics, accents or speech patterns of the race or nationality in question" (*Clifford v. Kentucky*, 7 SW 3d 375–76).

58. This concern is the center of the discussion in Eidsheim, "Measuring Race."

59. The assessments here in terms of gender are infinitely complex. Why do we leave unexamined the question of whether the voice might even be the sound of somebody imitating a female's voice? Or of a female voice impersonating the sound of a man, a child, an animal, and so on?

60. *Clifford v. Kentucky*, 7 SW 3d, 373.

61. This is how Ralph Nader, then an independent candidate for president, referred to Senator Barack Obama, then the presumed Democratic nominee, in an interview (Ralph Nader, "Nader: Obama Trying to 'Talk White' and 'Appeal to White Guilt,'" *Huffington Post*, July 7, 2008, accessed April 19, 2015, http://www.huffingtonpost.com/2008/06/25/nader-obama-trying-to-tal_n_109085.html).

62. Robert Fink, e-mail message, September 13, 2014.

63. The physicist Charles Ross explains that the term *acoustic shadow* describes a situa-

tion in which a given person normally would hear a sound but does not hear the sound. Ross breaks the causes into three acoustic conditions: sound absorption, wind shear, and temperature gradients. For technical details, see "Blending History with Physics: Acoustic Refraction," *The Physics Teacher*, 38 (2000): 208–9 (http://digitalcommons.longwood.edu/cgi/viewcontent.cgi?article=1028&context=chemphys_facpubs). The acoustic shadow has fascinated many researchers interested in explaining various degrees of communication during the American Civil War. See *Civil War Acoustic Shadows* (Shippensburg, PA: White Mane, 2001).

64. Michel Martin, "Pastor Jim Wallis Back to Being Political," *Tell Me More*, National Public Radio, April 12, 2013, accessed April 25, 2013, http://www.npr.org/2013/04/12/177032267/pastor-jim-wallis-back-to-being-political. Jim Wallis also writes about this in *On God's Side: What Religion Forgets and Politics Hasn't Learned about Serving the Common Good* (Grand Rapids, MI: Brazo, 2013).

65. At this point, the concepts of sound and music have been deflated and could be written as if erased (~~sound~~ and ~~music~~).

66. Annemarie Mol, *The Body Multiple: Ontology in Medical Practice* (Durham, NC: Duke University Press, 2002), 6.

67. According to Jacki Lyden, "the essay argues for the responsible, ethical treatment of nature, which became the core of modern conservationism" ("Remembering Aldo Leopold, Visionary Conservationist and Writer," *All Things Considered*, National Public Radio, March 10, 2013, accessed March 1, 2015, http://www.npr.org/2013/03/10/173949498/remembering-aldo-leopold-visionary-conservationist-and-writer). Aldo Leopold's *A Sand County Almanac: With Essays on Conservation* (New York: Oxford University Press, [1949] 1964) set the stage for the conservation movement and has continued to influence generations of conservation scientists.

68. Leopold, *A Sand County Almanac*, 204.

69. Leopold's original text is: "It's inconceivable to me that an ethical relationship to the land can exist without love, respect and admiration, and a high regard for its value" (ibid., 223).

70. The analogy between music and our physical environment continues: like music, land is a process and event when considered within a tectonic timeframe.

71. Considering "black music," Fred Moten also writes about the "*mater*iality of voices that the music represents" (*In the Break: The Aesthetics of the Black Radical Tradition* [Minneapolis: University of Minnesota Press, 2003], 39).

72. In the context of understanding ventriloquism, Steven Connor has already offered the definition of voice as "event." He writes: "my voice is not something that I merely have, or even something that I, if only in art, am. Rather, it is something that I do. A voice is not a condition, nor yet an attribute, but an event" (*Dumbstruck*, 4).

73. Adriana Cavarero beautifully addresses voice as relational between unique beings (*For More Than One Voice*).

Chapter 1. Music's Material Dependency

1. Jonathan Sterne, *The Audible Past: Cultural Origins of Sound Reproduction* (Durham, NC: Duke University Press, 2003), 137. As the central concern of chapters 3 and 5, this critique is fleshed out later in the book.

2. Two other young southern California musicians have found solace and inspiration in the bathroom. To read about Erin "Rin" Perey's and Roxanne "Rox" Ilano's recording in the bathroom for its acoustics and "because it retained a comforting fantasy of privacy, relaxation, and (to phrase it politely) release," see Karen Tongson, "Choral Vocality and Pop Fantasies of Collaboration," *Journal of Popular Music Studies* 23, no. 2 (2011): 229–34.

3. The project as a whole is titled the *Five Fathoms Opera Project*. The workshop version in which my students and I participated is called "Aquaopera #4/Los Angeles." Currently, there are two additional distinct performances under the *Five Fathoms Opera Project* umbrella, *Five Fathoms Deep My Father Lies* (with sound design by Snapper) and *You Who Will Emerge from the Flood*. The latter is a large-scale show with choir for which Andrew Infanti has composed some of the music. (To offer descriptions or readings of the entire *Five Fathoms Opera Project* is beyond the scope of this book.) At the beginning of *You Who Will Emerge from the Flood*, Snapper enters the pool area in a costume designed by Susan Matheson that resembles a beautiful, yet eerie, transformation of seaweed into dress. Two men, with swords longer than themselves, escort her. Later a full chorus enters the scene, and the drama unfolds by the side of the pool, in the water, on an underwater stage, and on a video projection (see figure 1.1). Metaphorically, this is an expression of human creativity that defies the end of time. Seen in this context, one way of reading *Five Fathoms Opera Project* is as a thorough undermining of the attitude exhibited by the minority of Christians who viewed Katrina as divine punishment for homosexual activity.

Snapper's underwater performances thus far include:

Aquaopera #1/Palm Springs, collaboration with Jeanine Oleson; pool at private residence in Palm Springs, CA; July 14, 2007.

Aquaopera #2/San Francisco, Shotwell Shack Series; bathroom collaboration with Jeanine Oleson; San Francisco, CA; July 20, 2007.

Five Fathoms Deep My Father Lies; P.S.1 Contemporary Art Center/MoMA, New York City; March 15, 2008.

Five Fathoms Deep My Father Lies; REDCAT Series at the Standard Hotel rooftop pool, Los Angeles, CA; November 12, 2008.

Five Fathoms Deep My Father Lies; Aksioma, Institute for Contemporary Arts; sound design by Pieter Snapper; participation by full chorus; Ljubljana, Slovenia; June 20, 2008.

Aquaopera #3/Los Angeles; Sea and Space Gallery, February 5, 2009.

You Who Will Emerge from the Flood, collaboration with Andrew Infanti, full chorus; Queer Up North Festival, Manchester, England; May 17, 2009.

Aquaopera #4/New York; artist residency, workshops, and performance; collaboration with Jeanine Oleson; Denniston Hill Residency, Woodridge, NY; August and September 2009.

Five Fathoms Deep My Father Lies, collaboration with Jeanine Oleson; Viva! Art+Action Festival, Montreal, QC; September 18, 2009.

You Who Will Emerge from the Flood, collaboration with Andrew Infanti; full chorus; TRAMA Festival de Artes Performativas, Porto, Portugal; October 10, 2009.

Aquaopera #5/Los Angeles; Standard Hotel; April 28, 2010.

You Who Will Emerge from the Flood, collaboration with Andrew Infanti, full chorus; Theatr Dramatyczny festival "Migracje," Warsaw, Poland; October 24, 2010.

You Who Will Emerge from the Flood, collaboration with Andrew Infanti, Robert Steinberger, full chorus; La Batie Festival du Geneve, Geneva, Switzerland; September 10, 2011.

4. Scott Timberg, "Influences: Opera Singer Juliana Snapper," *Los Angeles Times*, July 18, 2012.

5. Snapper conceptualizes the human body as a technology in the broad sense of its processes providing solutions to particular needs (breathing, blood flow, moving through space, and so on), interview with author, April 1, 2009.

6. This is related to the ethnomusicologist Ana María Ochoa Gautier's project about the voice in nineteenth-century Colombia, where she discusses "how a zoopolitics of the voice, that is the use of voice to define the boundaries between the human and the non-human, generated contested understandings of difference that in turn yielded different notions of the boundaries between music and sound" ("The Zoopolitics of the Voice in Colombia in the Nineteenth Century," the Eleventh Annual Robert Stevenson Lecture at University of California, Los Angeles, May 2, 2013). See also Ana María Ochoa Gautier, *Aurality: Knowledge and Listening in Nineteenth-Century Colombia* (Durham, NC: Duke University Press, 2014).

7. Both Z and Norderval use the programming language MAX/MSP.

8. For an extensive discussion of underwater music, including Snapper's work, see Stefan Helmreich, "An Anthropologist Underwater: Immersive Soundscapes, Submarine Cyborgs, and Transductive Ethnography," *American Ethnologist* 34, no. 4 (2007): 621–41.

9. For a discussion of some music dealing with water without immersion, see Douglas Kahn, *Noise, Water, Meat: A History of Sound in the Arts* (Cambridge, MA: MIT Press, 1999).

10. According to Kahn, Cage traces his use of water to much later. In 1937 at the University of California, Los Angeles, he accompanied water ballet swimmers and sought to find a way to cue them while under water (see ibid., 249).

11. As I work on the copyedits of these pages, I learn that a workshop version of the project "AquaSonic" (described as a "music concert performed entirely underwater") is taking place at the Åbne Scene in Aarhus, Denmark (April 30–May 1, 2015) and the premier is scheduled for September 2015. The ensemble consist of vocals and specially adjusted instruments such as the organ, violin, electromagenetic harp, chimes, and percussions. Accessed April 24, 2015. See http://www.aquasonic.dk/#.

12. Michel Redolfi, "Underwater Music Facts, Interview with Kevin McGuinness,

Walt Disney Studio Home Entertainment, Sept 2008," accessed April 2010, http://www.redolfi-music.com/eau/UnderwaterMusicFaqs.pdf.

13. Roland Barthes, "The Grain of the Voice," in Roland Barthes, *Image, Music, Text*, essays selected and translated by Stephen Heath (New York: Hill and Wang, 1977), 179–89. Keep in mind that the backdrop for these works is voice primarily conceived as a metaphor for textual authority, while the sounded voice is largely subsumed by logos and, by extension, is imbued with authenticity and identity. To offer just one example from feminist scholarship—as eloquently pointed out in Leslie C. Dunn and Nancy A. Jones, *Embodied Voices: Representing Female Vocality in Western Culture* (Cambridge: Cambridge University Press, 1994)—voice is associated with the effort to reclaim one's experience through writing or having voice.

14. Christopher Small, *Musicking: The Meanings of Performing and Listening* (Middletown, CT: Wesleyan University Press, 1998), 2.

15. Suzanne G. Cusick, "On Musical Performances of Gender and Sex," in *Audible Traces: Gender, Identity and Music*, ed. Elaine Barkin and Lydia Hamessley (Los Angeles: Carciofoli Verlagshaus, 1999), 30.

16. Carolyn Abbate, "Music—Drastic or Gnostic?," *Critical Inquiry* 30, no. 3 (2004): 535.

17. Steven Connor, "Edison's Teeth: Touching Hearing," in *Hearing Cultures: Essays on Sound, Listening, and Modernity*, ed. Veit Erlmann (Oxford: Berg, 2004), 153–72.

18. Jul Snapper, "Juliana Snapper Work.Excerpts," accessed April 21, 2015, https://vimeo.com/81368087.

19. For example, Westboro Baptist Church, which has received much attention and criticism for its "God Hates Fags" campaign and demonstrations, espouses the theory that God kills people through floods. The church's website includes the statement: "16,000,000,000—people that God killed in the flood" (Westboro Baptist Church, "God Hates Fags," accessed July 10, 2012. http://www.godhatesfags.com).

20. Vestinavesti, "Interview with Juliana Snapper," accessed April 1, 2010. http://www.youtube.com/watch?v=gqDz6UCT5ws&feature=PlayList&p=06C40068B70D736D&playnext_from=PL&playnext=1&index=5.

21. Quoted in Mojca Kumerdej, "Sopranistko Juliano Snapper Pod Vodo, Sirenine Podvodne Arije" [Soprano Juliano Snapper's underwater sirensong], *Delo*, June 28, 2008, 24–26. Translated by Luke Dunne and Juliana Snapper.

22. A fundamentalist Pentecostalist minister's son, Athey often talks about his childhood ability to speak in tongues. Amelia Jones, "Holy Body: Erotic Ethics in Ron Athey and Juliana Snapper's Judas Cradle," *TDR* 50, no. 1 (2006): 163, footnote 10.

23. The poem "To Those Born Later" can be found in Philip Thomson, "Brecht's Poetry," *The Cambridge Companion to Brecht*, ed. by Peter Thomason and Glendyr Sacks (Cambridge: Cambridge University Press, 2006), 212.

24. Quoted in Kumerdej, "Sopranistko Juliano Snapper Pod Vodo."

25. Jones, "Holy Body."

26. Ibid., 167.

27. For a collection of essays on the connection between sirens and music, see Linda

Phyllis Austern and Inna Naroditskaya, eds., *Music of the Sirens* (Bloomington: Indiana University Press, 2006).

28. Quoted in Karolina Sulej, "Setki ust na skórze," [A hundred paragraphs on the skin], *Wysokie Obcasy*, October 27, 2010. Translated by Karolina Sulej. Polish language version available here: http://www.wysokieobcasy.pl/wysokie-obcasy/1,53662,8547394,Setki_ust_na_skorze.html, accessed April 24, 2015.

29. Snapper's performances have thus far been staged in pools rather than in the ocean, and she therefore does not literally sing with whales and dolphins. For obvious reasons, singing underwater is much more risky than playing an instrument. Working with researchers from the Scripps Institute of Oceanography, Snapper is gradually developing a practice suited to the ocean. To learn about underwater sound pollution, see Elena McCarthy, *International Regulation of Underwater Sound: Establishing Rules and Standards to Address Ocean Noise Pollution* (Boston: Kluwer Academic, 2004). For a broader discussion of an explicit connection between the animal world, the artists treated in this book, and the way in which I suggest we theorize and approach definitions and analysis of music and voice, see chapter 5.

30. Quoted in Sulej, "Setki ust na skórze."

31. Snapper, unpublished manuscript. Schoenberg asked Marie Pappenheim, a young doctor in his circle, to write a libretto on a topic of her choosing for his opera, in which the music would be controlled by the structure of the unconscious. Judging from her libretto and from the social and professional circles she frequented in Vienna, it seems that Pappenheim was familiar with the period's psychoanalytical thought and literature. It is also interesting to note that Bertha Pappenheim—the woman identified as Anna O. with whom Joseph Breuer worked for a number of years, and whose treatment Breuer and Sigmund Freud chronicled—has been identified as being related to Marie Pappenheim. For Snapper, *Erwartung*'s connection to the hysterical women and the vocal diagnostic and subsequent application of the talking cure, is crucial. For an extensive discussion of the scholarship on this topic, see Alexander Carpenter, "Schoenberg's Vienna, Freud's Vienna: Re-Examining the Connections between the Monodrama *Erwartung* and the Early History of Psychoanalysis," *Musical Quarterly* 93, no. 1 (2010): 144–81.

32. For a thoughtful analysis of how people are judged to a significant degree by the sound of their voices as sane or insane, male or female, and so on, and of how insanity and femininity are two concepts bound together in the voice, see Anne Carson, "The Gender of Sound," in *Glass, Irony, and God*, introduction by Guy Davenport (New York: New Directions, 1995), 119–42.

33. Quoted in Sulej, "Setki ust na skórze."

34. With the term *hystericism*, Snapper also refers to what she describes as the "'rewired' feminine body, as it was first imagined by Freud and Breur" (e-mail message, August 13, 2013).

35. Ibid.

36. Quoted in Sulej, "Setki ust na skórze."

37. Juliana Snapper, interview with author, December 1, 2008.

38. Sound does not transmit efficiently from air to water, or from water to air. The traveling sound is reflected off the surface. When the sound incident takes place in air, it

reflects off the water surface, as off a heavy object. When the sound incident takes place in water, it also reflects back into the water because of the lack of continuation of the heaviness.

39. Steven Connor, "Michel Serres's Five Senses," 1999, accessed June 19, 2012, http://www.stevenconnor.com/5senses.htm.

40. For histories of underwater acoustics, see J. B. Hersey, "A Chronicle of Man's Use of Ocean Acoustics," *Oceanus* 20, no. 2 (1977): 8–21; Herman Medwin, *Sounds in the Sea: From Ocean Acoustics to Acoustical Oceanography* (Cambridge: Cambridge University Press, 2005); Herman Medwin and Clarence S. Clay, *Fundamentals of Acoustical Oceanography* (Boston: Academic, 1998); Susan Schlee, *The Edge of an Unfamiliar World: A History of Oceanography* (New York: Dutton, 1973); Robert J. Urick, *Principles of Underwater Sound* (New York: McGraw-Hill, 1983). For a current introduction to the theory of sound propagation in ocean, see, for example, L. M. Brekhovskikh and Yu. P. Lysanov, *Fundamentals of Ocean Acoustics*, 3rd ed. (New York: Springer-Verlag, 2003).

41. For more detailed information on the speed of sound in liquid, see, for example, Medwin, *Sounds in the Sea*.

42. The discussion regarding bone conduction is drawn from Jens Blauert, *Spatial Hearing: The Psychophysics of Human Sound Localization*, rev. ed. (Cambridge, MA: MIT Press, 1997), 191–93.

43. Ibid., 191–92. For positions on the hypothesis that bone conduction is relevant in normal spatial hearing, see H. Hecht, "Über die Lokalisation von Schallquellen" [On the localization of sound sources], *Naturwissenschafte* 11, no. 18 (1923): 338; A. Kreidl and S. Gatscher, "Über die Lokalisation von Schallquellen" [On the localization of sound sources], *Naturwissenschafte* 11, no. 18 (1923): 337–38; and H. A. Wilson and Charles S. Myers, "The Influence of Binaural Phase Differences in the Localization of Sound," *British Journal of Psychology* 2, no. 4 (1908): 363–85. For works skeptical of this hypothesis, see H. Banister, "A Further Note on the Phase Effect in the Localization of Sound," *British Journal of Psychology* 15, no. 1 (1924): 80–81; Jens Blauert, "Untersuchungen zum Richtungshören in der Medianebene bei fixiertem Kopf" [Investigations of directional hearing in the median plane with the head immobilized], PhD diss., Technische Hochschule, Aachen, 1969; H. Carsten, H. Salinger, and H. Hecht, "Zur Frage der Lokalisation von Schallquellen" [On the question of the localization of sound sources], *Naturwissenschaften* 10, no. 14 (1922): 329–30; H. Hecht, "Zur Frage der Lokalisation von Schallquellen" [On the question of the localization of sound sources], *Naturwissenschaften* 10, no. 14 (1922): 329–30; Hans Kietz, "Das räumliche Hören" [Spatial hearing], *Acta Acustica United with Acustica* 3, no. 2 (1953): 73–86.

44. Furthermore, Raymond M. Stanley and Bruce N. Walker have shown that "bone-conduction transducers offer a unique advantage for radio communication systems, allowing sound transmission while the ear canals remain open for access to environmental sounds, or plugged for blocking of environmental sounds" ("Intelligibility of Bone-Conducted Speech at Different Locations Compared to Air-Conducted Speech" [paper presented at the annual meeting of the Human Factors and Ergonomics Society, San Antonio, TX, October 19–23, 2009]). Interestingly, even in the condition of air, it is not only sound waves as transmitted through air that contribute to our understanding of

sound. Bryan Gick and Donald Gerrick have found that even air puffs reaching ankles can contribute to a listener's distinction between /b/ and /p/ ("Aero-Tactile Integration in Speech Perception," *Nature* 462, no. 7272 [2009]: 502–4).

45. Friedrich Kittler (*Gramophone, Film, Typewriter* [Stanford, CA: Stanford University Press, 1999]) posits that the stereo spatialization of cranial interiors became possible only with headphones.

46. Helmreich, "An Anthropologist Underwater," 624.

47. The late composer and sound artist Maryanne Amacher worked to expand our ways of hearing sound by purposefully using the physicality of the body beyond the eardrum for transmission. What she dubbed "direct sound"—sound being sent toward listeners, aiming for the eardrum—is the default listening mode, which we often imagine to be the only option. (Maryanne Amacher, "Maryanne Amacher in Conversation with Frank J. Oteri," April 16, 2004, Kingston, NY, videotaped by Randy Nordschowl transcribed by Molly Sheridan and Randy Nordschow, https://www.newmusicbox.org/assets/61/interview_amacher.pdf, accessed April 21, 2015.) In contrast, Amacher worked with frequencies distributed through space in such a way that the body itself would function as a speaker.

48. Snapper's team also had to deal with the challenging acoustics of swimming pool halls with hard surfaces and long reverberation time. In an effort to counteract this, they placed small speakers close together behind the audience at median ear level. The result was that the sound was transferred via microphone and speakers. Specifically, from the speakers, the sound waves were emitted to the audience from a fairly close distance, and therefore the sounds reflected from the walls did not affect their perception very much. In this way, the sound could be perceived as relatively sharp, despite the acoustics of the room. The effect, Snapper recounts, was a "kind of emersion." In the Geneva performance they were able to position the chorus right under the spectator seating, to, in Snapper's words, "sing up their arses, ha!" (e-mail message, August 13, 2013).

49. Michelle Duncan, "The Operatic Scandal of the Singing Body: Voice, Presence, Performativity," *Cambridge Opera Journal* 16, no. 3 (2004): 301.

50. Wayne Koestenbaum, *The Queen's Throat: Opera, Homosexuality, and the Mystery of Desire* (New York: Poseidon, 1993).

51. Thomas J. Csordas, "Embodiment as a Paradigm for Anthropology," *Ethos* 18, no. 1 (1990): 5.

52. Simone de Beauvoir, *The Second Sex* (New York: Vintage, 1974), 7.

53. As Tara Rodgers points out, the notion of the sound wave, on which much music analysis and thinking about music relies, is derived from a highly male-oriented perspective ("Toward a Feminist Epistemology of Sound: Refiguring Waves in Audio-Technical Discourse," in *Philosophy after Irigaray*, ed. Mary Rawlinson, Danae Mcleod, and Sara McNamara [Albany: State University of New York Press, forthcoming]).

54. Toril Moi, *What Is a Woman? and Other Essays* (Oxford: Oxford University Press, 1999).

55. It is interesting to observe that sharing the medium of water seems to be perceived as more material than sharing air, and to incite stronger visceral emotions about shared

materiality. This may be why policies regarding the sharing of the watery medium have often been enforced, such as pools segregated by gender and race.

56. Susan McClary, "Towards a Feminist Criticism of Music," *Canadian University Music Review* 10, no. 2 (1990): 11.

57. Thomas S. Kuhn, *The Structure of Scientific Revolutions* (Chicago: University of Chicago Press, 1965).

58. Ibid.

59. There are, of course, insightful exceptions where sound studies and certain areas of science and technology studies have moved beyond such perceived confines, which have made this work possible. See, for example the introductory chapter, note 7.

60. Sterne, *The Audible Past*, 15.

61. Franz Kafka, *The Complete Stories*, transl. Nahum Norbert Glatzer (New York: Schocken, 1995), 431. Additionally, in a close comparative dictional reading of the *Odyssey*'s sirens and the *Iliad*'s muses, Pietro Pucci suggests that the choice of στόμα to indicate the word *voice* in the *Illiad* "is used by the poet to indicate his own mouth as producing the song" and that "the Sirens, then, while inviting Odysseus to listen to their song, pick a usage that in the *Iliad* characterizes only the poet's and Odysseus' diction," suggesting that the words were mouthed by Odysseus ("The Song of the Sirens," *Arethusa* 12, no. 2 [1979]: 122).

62. It is crucial to acknowledge that Homer wrote of the sirens' *song*. By stressing song, I suspect he meant to suggest the sonorous, vibrational, and materially dependent component of voice rather than the linguistic component. Judith Peraino notes that it is crucial that song (rather than speech) is addressed in this myth. She further notes that Louis Althusser's notion of interpellation is a sort of story of the siren and posits that Althusser thinks through the voice in the register of speech and language. According to Peraino, while he reveals that a response to a call—turning around when a policeman yells "Hey, you there!"—is singular, Althusser fails to note that the story of hailing can end with more than one response: "The hailed individual turn[ing] around" need not only be a positive response. It can also end with the response "No, you've got the wrong guy" (Judith Peraino, *Listening to the Sirens: Musical Technologies of Queer Identity from Homer to Hedwig* [Berkeley: University of California Press, 2006], 3). Similarly, I would add that considering the sirens' song in terms of its textual content is the kind of thinking that contributes to the version of the story posited by Odysseus. Furthermore, I suggest that by taking the consequences of the power and work of the sonorous voice and its *song* seriously, we can interrogate the story further, as I do in this chapter.

63. David Tyson Copenhafer IV offers four possibilities: Odysseus hears and the sirens sing; Odysseus hears but the sirens do not sing; Odysseus does not hear and the sirens sing; and Odysseus does not hear and the sirens do not sing ("Invisible Ink: Philosophical and Literary Fictions of Music" [PhD diss., University of California, Berkeley, 2004], 88).

64. Mladen Dolar, *A Voice and Nothing More* (Cambridge, MA: MIT Press, 2006), 171–73.

65. Copenhafer, "Invisible Ink," 93.

66. Steve Goodman, *Sonic Warfare: Sound, Affect, and the Ecology of Fear* (Cambridge, MA: MIT Press, 2010), 191–92.

67. Max Horkheimer and Theodor W. Adorno, *Dialectic of Enlightenment: Philosophical Fragments*, trans. Gunzelin Schmid Noerr (Stanford, CA: Stanford University Press, 2002).

68. Ibid., 48. Friedrich Kittler stages his discussion of the sirens in direct opposition to Horkheimer and Adorno (*Musik und Mathematik I/1* [Munich: Fink], 2006). For an extensive discussion and context of Kittler's position, see Geoffrey Winthrop-Young, *Kittler and the Media* (Cambridge; Malden, MA: Polity Press, 2011), 82–119.

69. Lawrence Kramer, "'Longindyingcall': Of Music, Modernity, and the Sirens," in *Music of the Sirens*, ed. Linda Phyllis Austern and Inna Naroditskaya (Bloomington: Indiana University Press, 2006), 195.

70. Adriana Cavarero, *For More Than One Voice: Toward a Philosophy of Vocal Expression*, trans. Paul A. Kottman (Stanford, CA: Stanford University Press, 2005), 113.

71. Alexander Rehding is also interested in the "physiological effects of the siren songs on the body." By "debarrass[ing] music of the whole apparatus of art, of beauty and the philosophy that has been imposed on it," and instead considering "a kind of aesthetics in the Greek sense of *aeisthesis*, or perception," he hopes we can better inquire "into the sensuous parts of sound" ("Of Sirens Old and New," in *Oxford Handbook of Mobile Music Studies*, ed. Sumanth Gopinath and Jason Stanyek [Oxford: Oxford University Press, 2014], 2:82, 96).

72. Kafka, *The Complete Stories*, 431.

73. David B. Chamberlain, "The Fetal Sense: A Classical View," accessed April 21, 2015. https://birthpsychology.com/free-article/fetal-senses-classical-view. See also Peter G. Hepper and B. Sara Shahidullah, "Development of Fetal Hearing," *Archives of Disease in Childhood—Fetal and Neonatal Edition* 71, no. 2 (September 1, 1994): F81–F87; and Remy Pujol, Morello Lavigne-Rebillard, and Alain Uziel, "Development of the Human Cochlea," *Acta Oto-laryngologica* 111, no. s482 (1991): 7–13.

74. While I am aware that A has shifted over time, and that even orchestras around the world are now pulling the A "up" to create brilliance, I refer to this hertz measurement as the number to which we most commonly refer.

75. McClary, "Towards a Feminist Criticism of Music," 14.

Chapter 2. The Acoustic Mediation of Voice, Self, and Others

1. While I lay the groundwork here, I carry this project out in earnest in my book in progress, "Measuring Race: Listening to Vocal Timbre and Vocality in African-American Music" (unpublished manuscript).

2. For some historical background on concert halls and acoustic concerns, see Ian Appleton, *Buildings for the Performing Arts: A Design and Development Guide* (Boston: Butterworth Architecture, 1996); Leo Leroy Beranek, *Acoustic Measurements* (New York: John Wiley, 1949), *Concert Halls and Opera Houses: Music, Acoustics, and Architecture*, rev. ed. (New York: Springer, 2004), and *Music, Acoustics and Architecture* (New York: Wiley,

1962); Jens Blauert, *Spatial Hearing: The Psychophysics of Human Sound Localization*, rev. ed. (Cambridge, MA: MIT Press, 1997); Barry Blesser and Linda-Ruth Salter, "Ancient Acoustic Spaces," in *The Sound Studies Reader*, ed. Jonathan Sterne (New York: Routledge, 2012), 186–96; Michal Forsyth, *Buildings for Music: The Architect, the Musician, and the Listener from the Seventeenth Century to the Present Day* (Cambridge, MA: MIT Press, 1985); Michael Hammond, *Performing Architecture: Opera Houses, Theatres and Concert Halls for the Twenty-First Century* (London: Merrell, 2006); George C. Izenour, Vern Oliver Knudsen, and Robert B. Newman, *Theater Design* (New York: McGraw-Hill, 1977); Christopher J. Jaffe, *The Acoustics of Performance Halls: Spaces for Music from Carnegie Hall to the Hollywood Bowl* (New York: W. W. Norton, 2010); Chris van Uffelen, *Performance Architecture + Design* (Salenstein, Switzerland: Braun, 2010).

3. See Beranek, *Concert Halls and Opera Houses*; Forsyth, *Buildings for Music*; and Jeremy Montagu, *Origins and Development of Musical Instruments* (Lanham, MD: Scarecrow Press, 2007), 117.

4. This example is offered in Michael Hurd and John Borwick, "Concert Halls," in *The Oxford Companion to Music*, ed. Alison Latham (New York: Oxford University Press, 2002), 287. However, the first hall listed by Forsyth (*Buildings for Music*, 329) is the Eisenstadt Castle, which was completed in 1700. Forsyth provides two useful chronological tables of concert halls and opera houses, respectively (ibid., 329–33). Besides the building, year of completion, and architect, the table of concert halls also includes the hall volume in cubic feet and meters, number of seats, and reverberation time in seconds and at middle frequencies as the hall is filled with audience members (ibid., 329).

5. The quote and information in this paragraph are from Hurd and Borwick, "Concert Halls."

6. Beranek, *Concert Halls and Opera Houses*, 11.

7. By *inclusive* I do not mean that anyone could, or was encouraged to, attend. While the concerts were public, there were (and still are) restrictions in regard to concert attendance. However, the simple point I wish to make here is that, when moving from the private to the public realm, a larger segment of the population was eligible to attend concerts; thus the required number of seats increased.

8. Jaffe reports that the Christchurch Town Hall for Performing Arts, in New Zealand, was one of the earliest concert halls to offer surround seating. This was part of a trend that began in the 1970s toward what was perceived as more egalitarian seating (a greater number of seats closer to the orchestra) and more intimacy (with audiences closer to the orchestra). For additional reflections from an acoustician's point of view, see Jaffe, *The Acoustics of Performance Halls*, 95.

9. Beranek, *Concert Halls and Opera Houses*, 14.

10. Ibid., 14–15.

11. Ibid., 15.

12. Thurston Dart, *The Interpretation of Music* (London: Hutchinson's University Library, 1954), 56–57.

13. Robert Fink, "Unwrapping the Box: Gehry's Walt Disney Concert Hall as Postmodern Space" (paper presented at the annual meeting of the American Musicological Society, Philadelphia, PA, November 12–15, 2009), 11.

14. Ibid.

15. The term *pure sound* is used to describe a sound without harmonics (that is, a sound made up of a single frequency), such as the sound produced by a tuning fork. However, here I use the term more in alignment with Ari Kelman's description of the characters in Emily Ann Thompson's study (*The Soundscape of Modernity: Architectural Acoustics and the Culture of Listening in America, 1900–1933* [Cambridge, MA: MIT Press, 2002]), who "sought to create sound that could be isolated or abstracted from its context" and strove to gain complete control of the sound, despite the acoustic situation (Ari V. Kelman, "Rethinking the Soundscape: A Critical Genealogy of a Key Term in Sound Studies," *Senses and Society* 5, no. 2 [2010]: 225).

16. Clifford Siskin and William Warner, "This Is Enlightenment: An Invitation in the Form of an Argument," in *This Is Enlightenment*, ed. Clifford Siskin and William Warner (Chicago: University of Chicago Press, 2010), 1.

17. These reverberation times are reported by Leo Beranek in "Analysis of Sabine and Eyring Equations and Their Application to Concert Hall Audience and Chair Absorption," *Journal of the Acoustical Society of America* 120, no. 3 (2006): 1404. For additional detail on "audience absorption coefficient," see table II in the same publication. For further reading, please see Leo Beranek, "Subjective Rank-Ordering and Acoustical Measurements for Fifty-Eight Concert Halls," *Acta Acustica United with Acustica* 89, no. 3 (2003): 494–508.

18. Jaffe, *The Acoustics of Performance Halls*, 63. Beranek quotes, at length, a very interesting reflection on the acoustics of the major U.S. concert halls by the conductor James DePreist, in which DePreist recounts growing up with the sound of the Philadelphia Orchestra and realizing that varying acoustics occasionally gave rise to musicians' need to "compensate for the dryness of the hall" (quoted in Beranek, *Concert Halls and Opera Houses*, 6). DePreist goes on to compare playing the same repertoire with the Philadelphia Orchestra in Philadelphia's Academy of Music and Carnegie Hall.

19. However, while in early concert hall developments spatial concerns greatly influenced acoustics, Emily Thompson's broader investigation of acoustics and sound concerns—including noisy streets, echoic lecture halls, and radio broadcast booths—in the early twentieth century revealed that acoustic concerns had a greater impact on spatial design (*The Soundscape of Modernity*).

20. Beranek, *Concert Halls and Opera Houses*, 2, 35. Beranek is very careful to note that these qualities are those most often desired by audiences, musicians, conductors, and critics.

21. Ibid., 35. Due to his practical and theoretical work, Beranek is understood as the North American (and probably global) authority on concert hall acoustics. He also published the most influential texts on the topic. See Christopher Jaffe, Mark Holden, and Robert Lilkendey, "Consistency in Acoustic Design When Building Symphonic Performance Halls," accessed June 20, 2014, http://www.siebeinacoustic.com/publications/papers/1999%20-%20Consistency%20in%20Acoustic%20Design%20When%20Building.pdf.

22. It is my understanding that Beranek first published the list in 1962. See Beranek, *Music, Acoustics and Architecture*, 20–34, 62–71.

23. The terms below are derived from Beranek, *Concert Halls and Opera Houses*, 18–42.

24. Ibid., 19. Jaffe developed "The Architectural Acoustic Translation System," a comparison of musical, scientific, and architectural vocabulary, from Beranek's list. For example, what in "Musical Vocabulary" is described as "Warmth, low string balance," is understood in "Scientific Vocabulary" as "Arrival time of low frequency." Furthermore, in "Architectural vocabulary," it is explained as "geometry of the hall," "Volume to seating area ratio," "Absorption in the hall," "Containment of volume," and "Coupled volumes near the source." See Jaffe, *The Acoustics of Performance Halls*, 34.

25. In his book, Beranek also occasionally refers to churches and chamber music halls. However, he notes: "The acoustics of small rooms and broadcast studios, which typically have problems at the low frequencies, are not treated here" (*Concert Halls and Opera Houses*, 20). As Beranek recognizes, while different terms might be used, there seems to be a good deal of general consensus about both acoustic excellence and acoustics that serve the music.

26. Jaffe, *The Acoustics of Performance Halls*, 34.

27. Ibid.

28. Beranek also developed a shorter list entitled "Vocabulary of subjective attributes of musical-acoustic quality" that offers the "quality" such as "intimacy," "warmth," and "balance" in noun form and adjectival form. The "antithesis"—respectively "lack of intimacy," "lack of bass," and "imbalance"—is also given in noun form and adjectival form (*Music, Acoustics and Architecture*, 64).

29. Fink, "Unwrapping the Box," 5.

30. Izenour, Knudsen, and Newman, *Theater Design*.

31. The point here is that this way of experiencing sound, which is understood as the default and correct way, is only a single way among limitless possibilities.

32. These processes are attempting to combat the basic principles of sound. The German psychoacoustician Jens Blauert explains: "Sound events and auditory events are distinct in terms of time, space, and other attributes . . . ; that is, they occur only at particular times, at particular places, and with particular attributes. The concept of 'spatial hearing' acquires its meaning in this context. Since this concept implies that auditory events are inherently spatially distinct, it is in fact tautological; there is no 'nonspatial hearing.' Defined more narrowly, the concept of spatial hearing embraces the relationships between the localizations of auditory events and other parameters—particularly those of sound events, but also others such as those related to the physiology of the brain" (*Spatial Hearing*, 3). For a brief historical introduction to research on sound localization, see Frances S. Hickson and Valerie E. Newton, "Sound Localization: Part I," *Journal of Laryngology and Otology* 95, no. 1 (1981): 29–40.

33. The quote is from Fink, "Unwrapping the Box," and its context is explained below. Artificially created reverberation is, of course, part of some composers' arsenal. See, for example, the discussion of artificially created reverberation used as a compositional tool in Tarina Riikonen, "Shaken or Stirred—Virtual Reverberation Spaces and Transformative Gender Identities in Kaija Saariaho's *NoaNoa* (1992) for Flute and Electronics," *Organised Sound* 8, no. 1 (2003), 109–15. In this context, it is also worthwhile to mention all the work that research institutes such as the Institute for Research and Coordination

in Acoustics/Music (IRCAM), "one of the world's largest public research centers dedicated to both musical expression and scientific research," has put considerable resources into creating variable reverberation spaces and sound in motion that has made varying reverberation and sound placement within space viable compositional parameters. However, because of the highly specialized technology needed, many of these compositions are challenging, if not impossible, to re-create in other venues. See Institute for Research and Coordination in Acoustics/Music, "Who are We?," accessed April 1, 2014, http://www.ircam.fr/ircam.html?&L=1.

34. Anne Midgette, "CD Review: Meredith Monk, Ascending," *Washington Post*, May 27, 2012.

35. The correspondence between Monk and the National Endowment for the Arts regarding the question of funding category can be found in the Meredith Monk Archive, 1959–2006, New York Public Library, New York.

36. Sally Banes, "The Art of Meredith Monk," *Performing Arts Journal* 3, no. 1 (1978): 3–18; Deborah Jowitt, *Meredith Monk* (Baltimore, MD: Johns Hopkins University Press, 1997), "Monk and King: The Sixties Kids," in *Reinventing Dance in the 1960s: Everything Was Possible*, ed. Sally Banes, Andrea Harris, and Mikhail Baryshnikov (Madison: University of Wisconsin Press, 2003), 113–36, and *Time and the Dancing Image* (Berkeley: University of California Press, 1989); Carole Koenig, "Meredith Monk: Performer-Creator," TDR 20, no. 3 (1976): 51–66; Nancy Putnam Smithner, "Meredith Monk: Four Decades by Design and by Invention," TDR 49, no. 2 (2005): 93–118.

37. Babeth M. VanLoo, dir., *Inner Voice: Meredith Monk* (Hilversum, the Netherlands: House Foundation for the Arts, 2008), DVD.

38. Different terms are in circulation regarding these techniques, but Monk uses the term "interdisciplinary performer" to describe herself. See "Biography," Meredith Monk, accessed April 1, 2013, http://www.meredithmonk.org/about/bio.html.

39. VanLoo, *Inner Voice*.

40. This raises questions about the work and the archive as performance art. Recently, a strain of this broader conversation has been grounded in Marina Abramović's performance art and her Museum of Modern Art retrospective exhibition, "Marina Abramović: The Artist Is Present" (March 14–May 31, 2010), for which she trained a group of artists to perform restagings of pieces she had originally performed. See Matthew Akers and Jeff Dupre, dirs., *Marina Abramović: The Artist Is Present* (Chicago: Music Box Films, 2012), DVD; Holland Cotter, "700-Hour Silent Opera Reaches Finale at MoMA," *New York Times*, May 30, 2010; Sean O'Hagan, "Interview: Marina Abramović," *Observer*, October 2, 2010, accessed April 1, 2010, http://www.theguardian.com/artanddesign/2010/oct/03/interview-marina-abramovic-performance-artist; Judith Thurman, "Walking through Walls: Marina Abramović's Performance Art," *New Yorker*, March 8, 2010, accessed April 1, 2014, http://www.newyorker.com/magazine/2010/03/08/walking-through-walls.

41. VanLoo, *Inner Voice*.

42. Boosey and Hawkes publishes these thirty-five scores, including choral, piano, voice(s) and orchestra, and opera works.

43. VanLoo, *Inner Voice*.

44. Ibid.

45. Ibid.

46. Bonnie Marranca, "Performance and the Spiritual Life: Meredith Monk in Conversation with Bonnie Marranca," *Journal of Performance and Art* 31, no. 1 (2009): 16–36.

47. The quotes in this paragraph are from Krista Tippett, "Meredith Monk: Archaeologist of the Human Voice," *On Being*, American Public Media, February 16, 2012, accessed April 1, 2014, http://www.onbeing.org/program/meredith-monk039s-voice/transcript/1401.

48. Thompson, *The Soundscape of Modernity*, 3.

49. Ibid., 7. While Thompson specifically discussed the Radio City Music Hall, the work and sonorous ideals of these sound engineers is not isolated from the broader trend. Around the same time, the media critic Rudolf Arnheim advocated for electronic media's ability to explore the effects of the "pure sound" of aural communication. He held that "resonance is eliminated, out of a very proper feeling that the existence of the studio is not essential to the transmission and therefore has no place in the listener's consciousness" and that "the listener rather restricts himself to the reception of pure sound, which comes to him through the loudspeaker" (*Radio* [London: Faber and Faber, 1936], 142–43).

50. Max Horkheimer and Theodor W. Adorno, *Dialectic of Enlightenment: Philosophical Fragments*, trans. Gunzelin Schmid Noerr (Stanford, CA: Stanford University Press, 2002).

51. Fink ("Unwrapping the Box," 9) recalls that Gehry summarized the situation in a meeting this way: "[Nagata's] criteria says [sic] Berlin stinks." The acoustic in question is Hans Scharoun's 1962 design for the Berlin Philharmonic Hall, in which the seats are divided over terraces, as in a vineyard, rather than organized in parallel rows, front to back.

52. *Songs of Ascension* was performed on October 29, 30, and 31, and November 1 and 2, 2008, at the REDCAT Theater, Los Angeles.

53. "An Opera for Headphones" was the tagline used in *Invisible Cities* media coverage. See, for example, "Artbound Special Episode: 'Invisible Cities,'" KCET, January 4, 2015, accessed March 6, 2015, http://www.kcet.org/arts/artbound/counties/los-angeles/artbound-special-episode-invisible-cities.html and Sennheiser press release "Invisible Cities: An Opera for Headphones Receives Emmy® for KCET Documentary of Landmark Headphone Opera Production," July 30, 2014, accessed March 6, 2015, http://en-us.sennheiser.com/news-invisible-cities-an-opera-for-headphones-receives-emmy-for-kcet-documentary-of-landmark-headphone-opera-production. In his annual account of the city's cultural life, Mark Swed of the *Los Angeles Times* has twice noted Sharon's impact ("Best-of-2013: Music," *Los Angeles Times*, December 22, 2013, and "Classical Music: Faces to Watch 2012," *Los Angeles Times*, January 1, 2010).

54. This is especially interesting given the conspiracy theories about specific singers and their alleged microphone use, which is thought to weaken their voice and art (see Kai Harada, "Opera's Dirty Little Secret," *Live Design*, March 1, 2001, accessed April 1,

2012, http://livedesignonline.com/mag/operas-dirty-little-secret; Anthony Tommasini, "Wearing a Wire at the Opera, Secretly, of Course," *New York Times*, June 28, 2013). However, perhaps the high-definition live streaming of the Metropolitan Opera's performances has led us to conceptualize a viable connection between skilled operatic voices and live performances using microphones, as we have accepted recorded operatic voices mediated through microphones for decades.

A 2013 *New York Times* article was devoted to the anxiety of some opera audiences about the presence of a microphone on the opera singer's body. Tommasini writes, "As someone who cherishes classical music as an art form that glories in natural sound (while fully appreciating that many contemporary composers have used amplification in sonically alluring ways) I get nervous hearing Mr. Gelb [the Metropolitan Opera's general manager] talk of camouflaging wires on singer's bodies. And the Met has certainly kept this practice secret." In the same article, Jay David Saks, the sound designer for the Met's live broadcasts, responds to this perceived anxiety over keeping it natural, saying he prefers to "avoid wiring singers. 'For one thing, I don't get calls from people wanting to know why we do this sort of thing,' he said. 'It would be a lot easier without them. It ratchets up the complexity of my job.'" Toward the end of the article, Tommasini writes: "But Mr. Saks strongly rejected the idea that body microphones represent a more intrusive kind of amplification. 'I would bet everything I own that from listening to the broadcasts you could not tell which singers and which productions used body microphones,' he said. 'This story,' he said, referring to my pursuit of the matter, 'started from something someone saw, not from something someone heard'" ("Wearing a Wire at the Opera").

While the acoustic operatic voice is fetishized, the recorded operatic voice is not vilified in the same way that a live opera singer who might be using a microphone is. Indeed, one of the first commercially viable record companies, the Victor Talking Machine Company, forged a strong connection with Enrico Caruso, using him as the company's exclusive spokesperson (Richard Leppert, "Phonography and Operatic Fidelities (Regimes of Musical Listening, 1904–1929)" [paper presented at the University of Chicago Music Department's Colloquium Series, May 2, 2014]).

55. Wayne Koestenbaum (*The Queen's Throat: Opera, Homosexuality, and the Mystery of Desire* [New York: Poseidon, 1993]) and Michel Poizat (*The Angel's Cry: Beyond the Pleasure Principle in Opera*, trans. Alain Didier-Weill [Ithaca, NY: Cornell University Press, 1992]) describe specific articulations of this phenomenon, as observed through the opera fan's perspective.

56. Quoted in "Artbound Special Episode."

57. Fred Cummins, "Rhythm as an Affordance for the Entrainment of Movement," *Phonetica* 66, nos. 1–2 (2009): 15–28; Sylvain Rene Yves Louboutin, "Coordinated Group Musical Experience," *Google Patents*, US8521316 B2, filed March 31, 2010, issued August 27, 2013, accessed January 1, 2015, http://www.google.com/patents/US8521316.

58. Mark Swed has written about *Invisible Cities* in the context of site-specific work, comparing it to *Alter Banhof Video Walk*, a piece by the sound installation artist Janet Cardiff and her collaborator George Bures Miller that was presented at dOCUMENTA

13, June 9–September 12, 2012, in Kassel, Germany. See Mark Swed, "Moving Sound to Anywhere but the Concert Hall," December 28, 2013.

59. Quoted in "Artbound Special Episode."

60. The original text is: "There is a sense of emptiness that comes over us at evening" (Italo Calvino, *Invisible Cities*, trans. William Weaver [Orlando, FL: Harcourt Brace, 1972], 5).

61. Quoted in "Artbound Special Episode."

62. Ibid.

63. Calvino, *Invisible Cities*, 135.

64. Christopher Reynolds, "Union Station Bustles with Film Plots," *Los Angeles Times*, November 22, 2013.

65. For an illuminating and discourse-setting critique of the politics of downtown Los Angeles's development, see Mike Davis, *City of Quartz: Excavating the Future in Los Angeles* (London: Verso, 1990).

66. For a thorough review of literature on sound and space, see Georgina Born, *Music, Sound and Space: Transformations of Public and Private Experience* (Cambridge; New York: Cambridge University Press, 2013), especially the "Introduction," 1–69.

67. It is beyond the scope of the book's overarching argument to discuss the music in detail. I tackle this topic in an article in progress; see Nina Sun Eidsheim, "*Invisible Cities* at Union Station, Los Angeles 2014" (manuscript in progress). For more details on the music, see (and listen to the included excerpts) Christopher Cerrone, "Invisible Cities: Composing an Opera for Headphones," Artbound, October 22, 2013, accessed March 24, 2014, http://www.kcet.org/arts/artbound/counties/los-angeles/invisible-cities-opera-composer-christopher-cerrone-union-station-music.html.

68. The headphones are omnisonorous in that you can hear the same sound wherever you move within the wireless signal's range.

69. See Rick Altman, *Sound Theory, Sound Practice* (New York: Routledge, 1992).

70. I also see audience members begin to interact with Union Station patrons in a pedagogical or docent-like manner. Some audience members approach people who have stopped to observe some of the performers. I overhear audience members explaining the concept of an invisible opera in the station. At one point, I see an audience member being asked about the activity. Later, when I attend the performance *au naturel*, several headphone-wearing audience members offer to lend me their headphones. In one instance I let my son try the headphones for a minute.

71. In addition to the musical and performative elements, a camera crew and two projections add to the daily activity of the station. The camera crew feeds live images to one street-level projection, filling a wall, and can closely mirror the activity in that space. This projection shows a mix based on a live feed of various elements of the performance and activity in the space. Another subtitled projection offers a translated version of the libretto.

72. Don Ihde (*Listening and Voice: A Phenomenology of Sound* [Athens: Ohio University Press, 1976]) has pointed out that we can gauge the size of a physical space through sight and sound, but that sometimes contradictions (for example, mirrors or an anechoic

chamber) make this a challenging task. In this production, if a member of the audience listens only to the acoustics presented through the headphones, he or she is given a fictional mapping that bears no indicators of the physical placement and relationships between singers and in regard to singers' relationship to the physical space of the station.

73. Lisa Napoli, "The Drama of Humanity Unfolds in Union Station—oh, and an Opera, too," KCRW, *Which Way, LA?* October 18, 2013, accessed November 1, 3013, http://blogs.kcrw.com/whichwayla/2013/10/the-drama-of-humanity-unfolds-in-union-station-oh-and-an-opera-too.

74. There may be exceptions, such as sonic hallucinations, although I suspect that most of those also come with an acoustic character, not just pitch, timbre, duration, and so on.

75. Eugene Ormandy, unpaginated foreword to Leo Leroy Beranek, *Music, Acoustics and Architecture* (New York: Wiley, 1962).

76. Within this topic, it is interesting to consider moments when public concert culture undergoes a shift in values, and building and acoustic developments respond. Recent examples of responses to such changes can be seen in the development of the vineyard-style concert hall, which aims to distribute audiences in a more egalitarian layout that seeks to refrain from obviously class-based seating divisions. However, a thorough discussion of this phenomenon deserves more detailed consideration than the framework of this book allows.

Chapter 3. Music as Action

1. Richard Serra, *Boomerang*, 1974, accessed April 1, 2012, http://www.ubu.com/film/serra_boomerang.html.

2. *Boomerang* may bring to mind Alvin Lucier's 1969 composition *I Am Sitting in a Room*. However, there are some major differences in the vocal treatments. *I Am Sitting in a Room* calls for the reading of a written text, the recording of that reading and the recording of the sound of that recording, as it is sounded in a given architectural and acoustic structure. With each of the thirty-two called-for repetitions, the recording of the previous recording and its reverberation is sounded, while the live resonance of that recording is also sounded. Hence, each new recording records the playback of the previous recording and the live reverberation of the recording and the sound of the room. These instructions result in the continuous layering of the vocal recording and the given room's reverb. As it builds, the reverberation and layering smooth out the voice and move from recognizable sounds of language into nonlinguistic sounds. (Unlike the *Invisible Cities* production discussed in chapter 2, Lucier used the sound of the architectural space as an instrument itself.) In contrast to Lucier's work, *Boomerang* calls for improvised speech and does not utilize reverb (Nancy Holt, e-mail message, December 16, 2012). However, both pieces rely on processes that produce stutter. While *Boomerang* applies a process that prompts temporary stuttering as a by-product, *I Am Sitting in a Room* addresses and seeks to expose Lucier's own struggle with stuttering. As Lucier says

in the final sentence of the piece: "I regard this activity not so much as a demonstration of a physical fact, but more as a way to smooth out any irregularities my speech might have" (*I Am Sitting in a Room*).

3. This and subsequent quotes from *Boomerang* were transcribed from the recording available online (Serra, *Boomerang*). The line breaks and punctuations are mine; the elipses are used to indicate pauses.

4. The exact passage that the title is derived from is as follows: "The words become like things. I am throwing things into the world, and they are boomeranging back. They are boomeranging back."

5. Holt, e-mail message.

6. Ibid.

7. Rosalind Krauss, "Video: The Aesthetics of Narcissism," *October* vol. 1 (Spring 1976): 53.

8. Anne M. Wagner, "Performance, Video, and the Rhetoric of Presence," *October* vol. 91 (Winter 2000): 75. In a 1979 interview about his films, Richard Serra describes the process in this way: "There is a videotape called *Boomerang*, . . . with Nancy Holt, in which she is asked to respond to her own words which she hears through a feedback system. In a very detailed and clear way, she states what is happening to her as it is happening: her relation to herself as subject" (Annette Michelson, Richard Serra, and Clara Weyergraf, "The Films of Richard Serra: An Interview," *October* 10 [Autumn 1979]: 83).

9. Laura Malacart, "MUVE (Museum of Ventriloquial Objects): Reconfiguring Voice Agency in the Liminality of the Verbal and the Vocal" (PhD diss., University College London, 2011), 137. Malacart is an Italian native living in the United Kingdom. Her video work considers the accented English of immigrants and the ways in which their accents expose them to prejudice.

10. In addition, a topic that I do not discuss here—but that Holt specifically addresses at the end of the piece—is how the sensation of being in the physical and sensorial space of the video's creation (in the lighting, the tiny studio, and so forth) served as a metaphor for the experience of watching television. Holt ends the monologue with the following reflection: "Time in this isolated capsule of television experience is cut off from time as we usually experience it."

11. Jacques Derrida, *Of Grammatology*, trans. Gayatri Chakravorty Spivak (Baltimore: Johns Hopkins University Press, 1976), 281.

12. However, on the interweaving connection between the visual and music, sight and sound, Richard Leppert observes that music "connects to the *visible* human body, not only as the receiver of sound but also as its agent or producer. The human embodiment of music is central to any understanding of music's socio-cultural agency. The semantic content of music—its discursive 'argument'—is never solely about its sound and the act of hearing" ("Reading the Sonoric Landscape," in *The Sound Studies Reader*, ed. Jonathan Sterne [New York: Routledge, 2012], 409).

13. In fact, in searching YouTube for performances of *Boomerang*, I came across a version that uses Skype instead of video cameras (Perry Bard and Alejandro Jaramillo, "Boomerang: No Delay," February 8, 2011, accessed April 1, 2012, http://www.youtube.com

/watch?v=ERobX2de5DQ). This performance uses Holt's words as a script, so for the performers in this case, the effect of being exposed to their own voices is of course different from Holt's exposure to her own spontaneous reflections.

14. Kazutaka Kurihara and Koji Tsukada, "SpeechJammer: A System Utilizing Artificial Speech Disturbance with Delayed Auditory Feedback," accessed April 18, 2015, http://arxiv.org/abs/1202.6106. See also the information regarding the SpeechJammer on the two researchers' websites: Kazutaka Kurihara, "SpeechJammer," accessed March 7, 2015, https://sites.google.com/site/qurihara/top-english/speechjammer; Koji Tsukada, "SpeechJammer: Utilizing Artificial Speech Disturbance with Delayed Auditory Feedback," September 20, 2012, accessed March 7, http://mobiquitous.com/speechjammer-e.html.

15. For the 2012 prizewinners, see "Winners of the Ig Nobel Prize," Improbable Research, accessed March 7, 2015, www.improbable.com/ig/winners. The Ig Nobel Prize website describes the prizes as "honor[ing] achievements that make people laugh, and then think. The prizes are intended to celebrate the unusual, honor the imaginative — and spur people's interest in science, medicine and technology" ("About the Ig Nobel Prizes," Improbable Research, accessed March 7, 2015, http://www.improbable.com/ig/).

16. For instructions on creating homemade SpeechJammers, see, for example, Rituriel, "How to 'SpeechJam' with GarageBand," September 25, 2012, accessed July 16, 2013, http://www.youtube.com/watch?v=Aosw7GI2tFc; Phillip Garcia, "How to: Speech Jam / Jammer with Audacity," March 14, 2013, accessed July 16, 2013, https://www.youtube.com/watch?v=ho57bsII004; "Rhett and Link speech jammer (Garageband)," jim477, September 25, 2012, accessed April 23, 2015, https://www.youtube.com/watch?v=mXFsjqlo8rc. For an example of a speech-obstructing app, see "Researchers Build Directional 'SpeechJammer' Device," March 1, 2012, accessed April 23, 2015, http://www.electronista.com/articles/12/03/01/system.said.to.work.without.physical.discomfort/, a site known as "Electronista: gadgets for geeks," an article with the comforting sub-title, "System said to work without physical discomfort." According to the authors, "Japanese researchers have reportedly developed a 'SpeechJammer' device that is claimed to effectively block people from speaking. The device works using a psychological phenomenon that is said to make it difficult for someone to speak when their own words are played back to them with a short delay, wether [sic] they are reading from text or vocalizing an impromptu monologue." See also Martin Robinson, "A Mute Button for People? 'Speech Jamming Gun' That Stops People Talking by Freezing the Brain," March 2, 2012.

17. To watch friends testing the effect of the technology, see NickLaneTV, "Speech Jammer," April 7, 2013, accessed June 1, 2013, http://www.youtube.com/watch?v=Fzz7By8skLY; Noble Hower, "Speech Jammer with the Hower Family," December 28, 2012, accessed June 1, 2013, http://www.youtube.com/watch?v=X6nypeuC4j4; omfgwhatisthis shit, "Speech Jamminn!!," October 27, 2012, accessed June 1, 2013, http://www.youtube.com/watch?v=s_U_8x-Hiwc. For video blogs, see absuperman, "Speech Jammer Gun Review (Original)," May 31, 2013, accessed June 1, 2013, http://www.youtube.com/watch?v=oU9EGeMP5n4; Rhett and Link, "Speech Jammer Christmas Moments," Decem-

ber 9, 2012, accessed June 1, 2013, http://www.youtube.com/watch?v=YNHRsOdZ3ig. For an appearance on a talk show, see Ian Lortimer, dir., *QI XL*, "Jam, Jelly and Juice," March 12, 2013.

18. The software, Speech Zapper, created by Waron Sanguanwongwan.

19. Adriana Cavarero, *For More Than One Voice: Toward a Philosophy of Vocal Expression* (Stanford, CA: Stanford University Press, 2005), 29. For Mladen Dolar, however, "the non-articulate itself becomes a mode of the articulate; the presymbolic acquires its value only through opposition to the symbolic, and is thus itself laden with signification precisely by virtue of being non-signifying" (*A Voice and Nothing More* [Cambridge, MA: MIT Press, 2006], 24).

20. Serra, *Boomerang*.

21. Kurihara and Tsukada, "SpeechJammer."

22. For an insightful discussion on this, see Cavarero, *For More Than One Voice*.

23. For a summary of this position, see Jonathan Sterne, "Listening Is One of the Most Encultured Activities," *Weekend Edition*, NPR, September 14, 2002.

24. Cavarero, *For More Than One Voice*, 27, 11.

25. John Cage and Pierre Schaeffer are often cited for questioning the existence of silence and the notion that everyday sounds are nonmusical; see, for example, John Cage, *4'33": For Any Instrument or Combination of Instruments* (New York: Henmar Press: Edition Peters, 1993), and *Silence; Lectures and Writings*, 1st ed. (Middletown, CO: Wesleyan University Press, 1961); Pierre Schaeffer, *Traité Des Objets Musicaux* (Paris: Éditions du Seuil, 1966), and *In Search of a Concrete Music*, trans. Christine North and John Dack (Berkeley: University of California Press, 2012).

26. Music and action painting have been considered together before. See, for example, Jonathan Bernard, "Feldman's Painters," in *The New York Schools of Music and Visual Arts: John Cage, Morton Feldman, Edgard Varèse, Willem De Kooning, Jasper Johns, Robert Rauschenberg*, edited by Steven Johnson (New York: Routledge, 2002), 173–215; Mandy-Suzanne Wong, "Action, Composition—Morton Feldman and Physicality," 2009, accessed April 1, 2013, http://eventalaesthetics.net/sonauto/Wong%20-%20Feldman%20Turfan.pdf. There are particularly strong resonances here between my argument and Wong's work, as she discusses the possibility that Morton Feldman saw musical composition as a physical act, the point of which is the physical act itself.

27. Stan Brakhage and Joseph Cornell, dirs., *By Brakhage: An Anthology, Volume One* (Los Angeles: Criterion, 2010), DVD.

28. Robert Hughes, "An American Legend in Paris," *Time* 119, no. 5 (February 1982): 76.

29. The term is traditionally traced to a 1952 article, Harold Rosenberg, "The American Action Painters," in Barbara A. MacAdam, "Top Ten ARTnews Stories: 'Not a Picture but an Event,'" November 1, 2007, accessed January 1, 2013, http://www.artnews.com/2007/11/01/top-ten-artnews-stories-not-a-picture-but-an-event/.

30. Amelia Jones, *Performing the Body/Performing the Text* (London: Routledge, 1999), 52, 55.

31. A 2009 exhibition at the Pollock-Kranser House investigates the connection between Pollock and the Gutai Art Association and includes the Gutai journal in which

the link to Pollock is made in print (the group sent the journal to Pollock). Tetsuya Oshima, the curator who discovered Gutai's letter to Pollock among the Lee Krasner papers, speculates that the group's members were first exposed to Pollock's work in a 1951 exhibition in Japan. See Tetsuya Oshima, "'Dear Mr. Jackson Pollock': A Letter from Gutai," in *Under Each Other's Spell: Gutai and New York*, ed. Ming Tiampo (Stony Brook, NY: Stony Brook Foundation, Inc., 2009), 13–16. See also Benjamin Genocchio, "Painting with Hands and Feet," *New York Times*, August 21, 2009.

32. Kazuo Shiraga, "Challenging Mud," is an iconic piece from the period. It took place as an event during the group's first national exhibit, "The First Gutai Art Exhibition," outside Tokyo's Ōhara Hall in October 1955. See Namiko Kunimoto, "Shiraga Kazuo: The Hero and Concrete Violence," *Art History* 36, no. 1 (2013): 154–79. For a recollection of Saburo Murakami's paper screen piece, see Ming Tiampo, "Under Each Other's Spell: Gutai and New York," in *Under Each Other's Spell: Gutai and New York*, ed. Ming Tiampo (Stony Brook, NY: Stony Brook Foundation, Inc., 2009), 3–12. For a brief account of the Gutai Art Association's exchanges with American artists, see Yves-Alain Bois, "1955a," *Art since 1900: Modernism, Antimodernism, Postmodernism*, ed. Hal Foster, Rosalind Krauss, Yves-Alain Bois, and Benjamin H. D. Buchloh (London: Thames and Hudson, 2004), 2:411–13. Thanks to James Nisbet for educating me about the Gutai Art Association and their creative interaction with Pollock's work, and for the last reference in this note.

33. Jones, *Performing the Body/Performing the Text*, 53–63. See also Allan Kaprow, *Assemblage, Environments, Happenings* (New York: Harry N. Abrams, 1966). Thanks to James Nisbet for telling me about this, and for many interesting conversations on this and related topics. When studying Pollock's paintings, Kaprow saw them as leaping off the canvas and into the room; Pollock's work partly inspired Kaprow's art as performance and event. Not long after studying Pollock's paintings, Kaprow turned to performance scores similar to those created by composers of his time, including John Cage, Morten Feldman, and La Monte Young. Thus there are interesting reverberations between the so-called visual and sonic arts.

34. Mladen, *A Voice and Nothing More*, 24.

35. For examples of approaches to embodied analysis, see George Fisher and Judy Lochhead, "Analyzing from the Body," *Theory and Practice* 27 (2002): 37–68; Gascia Ouzounian, "Embodied Sound: Aural Architectures and the Body," *Contemporary Music Review* 25, no. 1 (2006): 69–79; Stacey Sewell, "Listening Inside Out: Notes on an Embodied Analysis," *Performance Research* 15, no. 3 (2010): 60–65.

36. In *Boomerang* Holt reflects that the situation "puts a distance between the words and their apprehension—their comprehension," a situation that is "like a mirror-reflection . . . so that I am surrounded by me and my mind surrounds me . . . there is no escape." In considering this passage, Krauss writes that Holt "describes and enacts" a "prison . . . from which there is no escape" ("Video," 53).

37. Together with Elodie Blanchard, I commissioned the *Noisy Clothes Suite*, which consists of a piece of mine and two pieces by Jonathan Moritz and Nick Chase respectively. Following the *Noisy Clothes* concept—action yields sound—these compositions were instructions for movement rather than descriptions of desired sonic results.

Instead of demanding the production of a signifying or ideal sound, the project invited participants to ask: "If I move my body in this way, what kind of sound will I make?" While we never worked according to an ideal of precision, which Pollock sometimes did and which suggests some dependence on preconceptions, *Noisy Clothes* echoed the more experimental aspects of Pollock's action painting. Through this project, we implicitly asked: are we capable of making a sound that is not just produced or understood as a reproduction of, or in a dynamic relation to, signifying sounds?

38. Clothes in the fashion designer Iris van Herpen's September 2013 collection "[create] their own 'music.'" The designer calls the audio waves that are "activated electronically by touch" "embossed sound." Quoted in Suzy Menkes, "The Sound of Invention," *New York Times*, October 1, 2013.

39. *Noisy Clothes* was performed at the California Institute of the Arts, Valencia, CA, May 16, 2000, and at the EARJAM festival, Los Angeles, CA, May 21, 2000.

40. Harold Rosenberg, *The Tradition of the New* (New York: Da Capo, 1994), 25.

41. The instruments were created from denim, elastene, paper, plastic, metal, and consumer electronics. The noisy clothes were designed using material categories or family groups within which interaction yielded sound. For example, the costume on the left hand side in figure 3.1 was made in part with and represented the Velcro family group. Each costume was made entirely of either the hooked or the looped side of the Velcro; therefore, to create sound, a costumed performer would have to physically connect to and disconnect from another performer's costume. The two remaining costumes in figure 3.1 included electronic components that created feedback sound. This couple's sound was also relational but, compared with the Velcro family, the sound resulting from their relationality was dispersed over a wider space and was created not only through literal connection and disconnection. Figures 3.3–3.4 are just examples of the idiosyncratic costumes that comprised the "instruments" of the Noisy Clothes Orchestra.

42. One important factor was that the participants who were professional musicians did not consider this performance project part of their regular music practice. Among our goals was isolating the project from performance contexts in which musicians felt that their reputations were at stake, because in such contexts external frames inhibit listening and performance.

43. Additionally, by framing rehearsals as social events, we sought to dispense with any notions of differences in skill between musicians and nonmusicians, helping the performers approach the project as a form of play.

44. Michel Foucault, *Discipline and Punish: The Birth of the Prison*, trans. Alan Sheridan (New York: Random House, 1995), 201.

45. My usage of the terms *normal* and *pathological* follows the work of Georges Canguilhem, who shows that the relationship between them is directed by the epistemological politics of difference (*On the Normal and the Pathological*, trans. Carolyn R. Fawcett [Dordrecht, Holland: Reidel Publishing Company, 1978]). Thanks to Jonathan Sterne for suggesting this framework.

46. Raymond Knapp and Mitchell Morris, "Singing," in *The Oxford Handbook of the American Musical*, ed. Raymond Knapp, Mitchell Morris, and Stacy Ellen Wolf (New York: Oxford University Press, 2011), 320–34.

47. Stephen Pennington's research on transgender voices also poignantly addresses how vocal sonic patterns are male- and female-coded, and how members of the transgender community translate elements of these codes into actionable exercises that offer the vocalizer expansion of the gender-coded vocal actions available to them ("Transgender Passing Guides and the Vocal Performance of Gender and Sexuality," in *The Oxford Handbook of Queerness and Music*, ed. Fred Maus and Sheila Whiteley [New York: Oxford University Press, forthcoming]).

48. David Sudnow, *Ways of the Hand: A Rewritten Account*, with introduction by Hubert L. Dreyfus (Cambridge, MA: MIT Press, 2001). Related is James Q. Davies's recent study of the "voices of virtuoso singers and the hands of virtuoso pianists" in London and Paris in the 1830s. Davies wishes to "assume an avowedly *realist* stance and ask how bodies are acquired as they are heard, trained, and performed." In other words, he asks: "How does music act in the cultivation of bodies?" (*Romantic Anatomies of Performance* [Berkeley: University of California Press, 2014], 2).

49. I wish to stress that with this piece I am not glorifying vocality without vocal cords. I recognize that people whose larynxes have been removed or who face challenges in the use of the larynx are able to choose whether or not to engage the vocal cords. However, I believe that thinking creatively about the function of the vocal cords can offer broader options for research on extended vocalization, or sources of vocal signals. See, for example, Rupal Patel and Anna Roden, "Intelligibility and Attitudes toward a Speech Synthesizer Vocoded Using Dysarthric Vocalizations," *Journal of Medical Speech-Language Pathology* 16, no. 4 (2008): 243–51.

50. Tia DeNora asks, rhetorically, whether the human body is the first musical instrument. She responds that humans' and other animals' bodies "possess sonic features" that "operate at several interconnected levels," including metabolic features, through contact with other material objects, body sounds like grunting or slurping, and vocal sounds ("Interlude 54. Two or More Forms of Music," in *International Handbook of Research in Arts Education*, ed. Liora Bresler [Dordrecht, the Netherlands: Springer, 2007], 800).

51. This definition would not overlap with linguistic and medical understandings of voice, the sound signal emitted from the vocal folds. (For example, the shaping of that signal by articulation is within the framework considered as manipulation of the voice signal.) Aristotle, however, defines voice sonically, anatomically, and in terms of the ability to make meaning: "Hence voice consists in the impact of the inspired air upon what is called the windpipe under the agency of the soul in those parts. For, as we have said, not every sound made by a living creature is a voice (for one can make a sound even with the tongue, or as in coughing), but that which even causes the impact, must have a soul, and use some imagination; for the voice is a sound which means something" (*On the Soul*, 420b.25–30; in Aristotle, *The Complete Works of Aristotle*, ed. Jonathan Barnes, trans. D'Arcy Wentworth Thompson [Princeton, NJ: Princeton University Press, 1984]). Aristotle brings this up in the context of discussing the anatomy of different species and the question of whether or not they have voice and/or language. For example, Aristotle draws a line: "It is clear also why fish are dumb [mute]; it is because they have no throat. They have not this organ because they do not take in air or breathe" (ibid., 421a.5). However, Aristotle does not equate language with meaning making: "Voice and sound are

different from one another; and language differs from voice and sound . . . language is the articulation of vocal sounds by the instrumentality of the tongue. Thus, the voice and larynx can emit vocal or vowel sounds; the tongue and the lips make non-vocal sounds or consonant sounds; and out of these vocal and non-vocal sounds language is composed" (*History of Animals*, 535a.27–535b.4; in Aristotle, *The Complete Works of Aristotle*).

52. Although I do not discuss it here, a significant aspect of the piece is that the corporeal vocal part—while not being registered, or activated, by the vocal cords—is registered by the biosensors. Affixed to a singer's body, these sensors provide data from which a programmer and a visual artist generate sounds and images. Thus, the singing process is measured with microsensors and presented to an audience via a prosthetic instrument that, in place of the vocal cords, sonifies and visualizes—or, if you will, transduces—the body's activity. The vocal cords, like the sensors and the data the monitor is set up to read, are thus not in fact the voice, but rather one of many possible transmissions or transductions of singing.

53. Björn Vickhoff and coauthors have suggested that "music structures determined heart rate variability of singers" ("Music Determines Heart Rate Variability of Singers," *Frontiers in Psychology* 4 [2013]: 1), while Victor Müller and Ulman Linderberger have shown that the cardiac and respiratory patterns are synchronized between members of a choir while they are singing ("Cardiac and Respiratory Patterns Synchronize between Persons During Choir Singing," *PLoS ONE* 6, no. 9 [2011]: e24893). Furthermore, experiments suggest that music induces changes in galvanic skin response (Daniel G. Craig, "An Exploratory Study of Physiological Changes During 'Chills' Induced by Music," *Musicae Scientiae* 9, no. 2 [2005]: 273–87), temperature (Chen-Gia Tsai, Rong-Shan Chen, and Tzung-Shian Tsai, "The Arousing and Cathartic Effects of Popular Heartbreak Songs as Revealed in the Physiological Responses of Listeners," *Musicae Scientiae* [2014]: 1–13), heart rate and systolic blood pressure (Wendy E. J. Knight and Nikki S. Rickard, "Relaxing Music Prevents Stress-Induced Increases in Subjective Anxiety, Systolic Blood Pressure, and Heart Rate in Healthy Males and Females," *Journal of Music Therapy* 38, no. 4 [2001]: 254–72), and dopamine release (Vinod Menon and Daniel J. Levitin, "The Rewards of Music Listening: Response and Physiological Connectivity of the Mesolimbic System," *NeuroImage* 28, no. 1 [2005]: 175–84).

54. DeNora goes on to say: "And these physiological manifestations, perceived by particular actors and within particular action scenarios, may come to serve as markers of status (bodily capital, positive and negative) and/or occasion (such as when we consider a situation to be fraught with tension or social discomfort)" ("Interlude 54," 802).

55. See Arthur H. Benade, *Fundamentals of Musical Acoustics* (New York: Dover, 1990), 360–75. Albert Einstein stated: "*If a body gives off the energy L in the form of radiation, its mass diminishes by L/c^2*" ("Does the Inertia of a Body Depend upon Its Energy Content?," in *The Collected Papers of Albert Einstein*, vol. 2, *The Swiss Years: Writings, 1900–1909*, trans. Anna Beck [Princeton NJ: Princeton University Press, 1989], 174; emphasis in the original).

56. The piece and its score went through multiple iterations. Six years elapsed between the first sketches and the version that, to me, best represented and conveyed instructions for the expression of voice as action. Each of *Body Music*'s ten breathing

stances is notated with a "trigger" name that briefly describes the stance and the breathing. In the excerpt above, for example, the first instruction consists of the words "suck short," indicating a quick, energetic inhalation through pursed lips. Each adapted "note," whether it is an arrow or an open circle, indicates events of inhalation (upward arrow) or exhalation (downward arrow). Rests (𝄾), which here indicate held breaths, are treated as pauses with relative durations. A tiny open circle or a plus sign indicate, respectively, an open or closed mouth, which results in the direction of air through the mouth or nose, respectively. The markings below the notes, such as /pa/ and /ps/, suggest adaptations to the position of the mouth that may create consonants or vowels. For example, /p/ would indicate loosely closed lips that opened to allow air to escape, giving rise to a sound resembling the pronunciation of the letter *p*.

57. Opera, though, is very much about the beauty of the sonorous (vocal cord) voice. What happens to opera when the sonorous voice takes a back seat to the body's hidden timbres? We may alternatively describe the process of singing as the alternation of inhalations and exhalations—and these actions are precisely the material of *Body Music*.

58. In the attempt to move away from the definition of *voice* as logos, I recognize that the term *vocabulary* might seem counterproductive. I invoke it deliberately to refer to the field of dance, in which bodily movements are referred to as *vocabulary*.

59. See Clifford Geertz, *The Interpretation of Cultures* (New York: Basic, 1973), 6.

60. Derrida, *Of Grammatology*, 44.

61. Ferdinand de Saussure, *Course in General Linguistics*, trans. Wade Baskin (New York: McGraw-Hill Book Company, 1966), 68.

62. Shane Butler, *The Ancient Phonograph* (New York: Zone Books, 2015), 31–58.

63. For example, Paul Iverson and coauthors recount how early language experiences can play a role in the ability or inability to acquire nonnative phonemes during adulthood ("A Perceptual Interference Account of Acquisition Difficulties for Non-Native Phonemes," *Cognition* 87, no. 1 [2003]: B47–B57). Their hypothesis states that early language experience helps shape sensitivity to specific acoustic cues such as, for example, the difference between /l/ and /r/. In addition, Christina M. Esposito shows that both Spanish and English speakers are inattentive to or unable to identify differences when exposed to contrastive vocal qualities not present in their native languages ("The Effects of Linguistic Experience on the Perception of Phonation," *Journal of Phonetics* 38, no. 2 [2010]: 306–16). Thanks Patricia Keating for pointing me to these references.

64. Saussure, *Course in General Linguistics*, 120.

65. Ibid., 120.

66. Jacques Derrida, *Positions*, trans. Alan Bass (Chicago: University of Chicago Press, 1981), 21.

67. I am very grateful to Shane Butler for extended conversations about writing, speech, and voice during the fall of 2012, conversations that were instrumental in helping me put these pieces together.

68. Roman Jakobson, "Why 'Mama' and 'Papa'?," in Roman Jakobson, *Studies on Child Language and Aphasia* (the Hague, the Netherlands: Mouton, 1971), 21–29.

69. Ibid., 22 (emphasis added).

70. I use the term *interpellation* loosely here, assuming a parallel between the situation

in which the police officer yells "you" and someone who feels guilty responds and the scenario in which the father, wishing to respond to his self-identity as a parent, overlays intentionality onto the baby's random sounds, hence interpellating himself.

71. Conversation with Shane Butler, December 4, 2014. Ann M. MacLarnon and Gwen P. Hewitt have discussed human breathing with the understanding that it has evolved differently from the breathing of nonhuman primates precisely because human breathing is modified for the fine respiratory control necessary for speech ("The Evolution of Human Speech: The Role of Advanced Breathing Control," *American Journal of Physical Anthropology* 109, no. 3 [1999]: 341–63). See also Jody Kreiman and Diana Sidtis, *Foundations of Voice Studies: An Interdisciplinary Approach to Voice Production and Perception* (Malden, MA: Wiley-Blackwell, 2011), 25–71.

72. For a thorough reading and theorizing of the primary Greek and Latin words for *voice*—*phōnē* and *vox*—please see Shane Butler, "What was the Voice?," unpublished manuscript. Their complexity, he notes, "begins to emerge when we consider what *else* they designate." Given the range of meaning these two words can include, we might as well ask "What was the (ancient) voice *not*?," he notes.

73. Leo Treitler, "The 'Unwritten' and 'Written' Transmission of Medieval Chant and the Start-Up of Musical Notation," *Journal of Musicology* 10, no. 2 (1992): 131–91, and "The Early History of Music Writing in the West," *Journal of the American Musicological Society* 35, no. 2 (1982): 237–79.

74. Thanks to Emma Dillon for explaining this to me in an e-mail communication, August 15, 2014.

75. Emma Dillon, *The Sense of Sound: Musical Meaning in France, 1260–1330* (New York: Oxford University Press, 2012), 25.

76. Ibid., 26.

77. Ibid., 38.

78. Ibid.

79. Augustine, *Confessions*, Book X, ch. 33 (§§ 49 and 50), in Augustine, *St. Augustine: Confessions*, trans. Henry Chadwick (Oxford: Oxford University Press, 1991), 207–8. About this passage, Dillon writes: "Augustine here expresses a fundamental tension between the sensuality of singing—the 'pleasure of the ear' ('voluptates aurium')—and the 'truth' or 'meaning' of the words sung (in this case, Psalms), expressed in the second part of the passage as a contrast between 'song' and 'meaning' ('cantus, quam res') and echoing a binary inherent in the unit of 'vox.' But Augustine also pinpoints a second important dynamic of that gap between sense and sound: the possibility that departure from 'res' is potentially revelatory, softening the listener for devotion—for feelings greater than the normal or everyday. At the same time, it is also potentially sinful, so much that he wishes he had never heard the singer" (*The Sense of Sound*, 40).

80. Bruce W. Holsinger, *Music, Body, and Desire in Medieval Culture: Hildegard of Bingen to Chaucer* (Stanford, CA: Stanford University Press, 2001), 7.

81. Dillon, *The Sense of Sound*, 41.

82. Ibid., 41.

83. Holsinger, *Music, Body, and Desire in Medieval Culture*, 1.

84. Ibid., 2.

85. Ibid., 8.

86. Jonathan D. Culler, *On Deconstruction: Theory and Criticism after Structuralism* (Ithaca, NY: Cornell University Press, 2007), 103.

87. Derrida, *Of Grammatology*, 281.

88. Ibid., 200.

89. Roland Barthes, "The Grain of the Voice," in *Image, Music, Text*, essays selected and translated by Stephen Heath (New York: Hill and Wang, 1977), 179–89.

90. There is a strong analogy here to Amy Cuddy's research on body language and the way in which it not only signals to the world around the person but also transforms the person who assumes those physical positions. For example, open posture leads to increases in testosterone levels and decreases in cortisol levels, which in turn produces a sense of power. In contrast, closed positions lead to decreased testosterone levels and increased cortisol levels. See Dana R. Carney, Amy J. C. Cuddy, and Andy J. Yap, "Power Posing: Brief Nonverbal Displays Affect Neuroendocrine Levels and Risk Tolerance," *Psychological Science* 21, no. 10 (2012): 1363–68.

91. This is a position similar to Antonio R. Damasio's regarding the notion of self from a neuroscientific point of view (*Descartes' Error: Emotion, Reason, and the Human Brain* [New York: G. P. Putnam, 1994] and *The Feeling of What Happens: Body and Emotion in the Making of Consciousness* [New York: Harcourt Brace, 1999]). It also echoes John Shepherd and Peter Wicke's notion of the sonic saddle (*Music and Cultural Theory* [Malden, MA: Blackwell], 159–68), borrowed from Victor Zuckerkandl, which is summarized by the sociologist Tia DeNora as "the tactile dimension of sound as it is presented to its hearers in a continually unfolding present" (*After Adorno: Rethinking Music Sociology* [Cambridge: Cambridge University Press, 2003], 103). The sonic saddle describes the way in which music has the capacity to, for example, bring us towards a climax unfolding over time, yet cannot solely be defined or explained in syntactical terms or temporal form. And it echoes DeNora's work (*Music in Everyday Life* [Cambridge: Cambridge University Press, 2000]) on the ways in which we use music in everyday life to curate an inner emotional life and the work of the ethnomusicologist Judith O. Becker (*Deep Listeners: Music, Emotion, and Trancing* [Bloomington: Indiana University Press, 2004]) on music's role in the transformation of the body that allows it to act as a vehicle for spiritual mediums.

92. Jennifer Stoever-Ackerman has used the term "the listening ear" "to describe how listening functions as an embodied cultural process that echoes and shapes one's orientation to power and one's posture toward the world" ("Splicing the Sonic Color-Line: Tony Schwartz Remixes Postwar Nueva York," *Social Text* 28, no. 1 [2010]: 64).

93. This position assumes an expanded notion of hearing, which is discussed in chapter 4.

94. Theresa Griffiths, dir., *Jackson Pollock: Love and Death on Long Island* (London: BBC, 1999), accessed April 1, 2013, http://www.bbc.co.uk/programmes/b0074km1.

95. Culler, *On Deconstruction*, 103.

Chapter 4. All Voice, All Ears

1. For two penetrating analyses of Shakespeare's balcony scene, which also were some of the foundational inspirations for this book, see Adriana Cavarero, *For More Than One Voice: Toward a Philosophy of Vocal Expression* (Stanford, CA: Stanford University Press, 2005), viii–xxi, 236–41; and Paul A. Kottman, *A Politics of the Scene* (Stanford, CA: Stanford University Press, 2008), 166–84.

2. Lydia Goehr, *The Imaginary Museum of Musical Works: An Essay in the Philosophy of Music* (Oxford: Clarendon Press of Oxford University Press, 1992).

3. Leo Treitler, "The 'Unwritten' and 'Written' Transmission of Medieval Chant and the Start-Up of Musical Notation," *Journal of Musicology* 10, no. 2 (1992): 131–91. Also, as discussed in chapter 3, Bruce W. Holsinger (*Music, Body, and Desire in Medieval Culture: Hildegard of Bingen to Chaucer* [Stanford, CA: Stanford University Press, 2001]) and Emma Dillon (*The Sense of Sound: Musical Meaning in France, 1260–1330* [New York: Oxford University Press, 2012]) offer useful examples of how musical analysis may be guided by contemporary contexts and relational dynamics.

4. Gary Tomlinson's "Musical Pasts and Postmodern Musicologies: A Response to Lawrence Kramer" (*Current Musicology* 53 [1993]: 18–24) was part of an extended exchange with Laurence Kramer, responding to the latter's "The Musicology of the Future" (*repercussions* 1, no. 1 [1992]: 5–18). Kramer responded to Tomlinson ("Music Criticism and the Postmodernist Turn: In Contrary Motion with Gary Tomlinson," *Current Musicology* 53 [1993]: 25–35), and Tomlinson responded to that ("Tomlinson Responds," *Current Musicology* 53 [1993]: 36–40).

5. Robert O. Gjerdingen and David Perrott, "Scanning the Dial: The Rapid Recognition of Music Genres," *Journal of New Music Research* 37, no. 2 (2008): 93–100.

6. Goehr, *The Imaginary Museum of Musical Works*, 89.

7. Jonathan Sterne, *The Audible Past: Cultural Origins of Sound Reproduction* (Durham, NC: Duke University Press, 2003), 21.

8. Ibid.

9. Amanda J. Weidman, *Singing the Classical, Voicing the Modern: The Postcolonial Politics of Music in South India* (Durham, NC: Duke University Press, 2006), 12.

10. The following vocal pedagogical texts are only a few examples of approaches that draw on both scientific and metaphorical approaches: Enrico Delle Sedie, *A Complete Method of Singing: A Theoretical and Practical Treatise on the Art of Singing* (New York: George Schirmer, 1894); William Johnstone-Douglas, "The Teaching of Jean de Reszke," in *Historical Vocal Pedagogy Classics*, ed. Berton Coffin (Lanham, MD: Scarecrow, 1989), 104–11; Lilli Lehmann and Richard Aldrich, *How to Sing*, rev. ed. (New York: Macmillan, 1914); Emma Seiler, *The Voice in Singing*, trans. William Henry Furness (Philadelphia: J. B. Lippincott, 1879); Julius Stockhausen, *A Method of Singing* (London: Novello, 1884); Herbert Witherspoon, *Singing, a Treatise for Teachers and Students* (New York: George Schirmer, 1925).

11. Both the institutional structure of the conservatory and its very existence are historically founded in a shift away from professional musicians, connoisseurs at court and court musicians, and religious services and their musical servants. The role of music

shifted from utilitarian to an artistic and potentially transcendent experience, and musicians' roles also gradually shifted from craftsmen and servants to expressive artists. This shift is also intimately connected to the formation and growth of the middle class, the Industrial Revolution, and the notion of autonomous art. For succinct summaries of this history and an assessment of its influence on music analysis, theory, and criticism, see, for example, Robin Moore, "The Decline of Improvisation in Western Art Music: An Interpretation of Change," *International Review of the Aesthetics and Sociology of Music* 23, no. 1 (1992): 61–84; Janet Wolff, "Foreword: The Ideology of Autonomous Art," in *Music and Society: The Politics of Composition, Performance and Reception*, ed. Richard Leppert and Susan McClary (Cambridge: Cambridge University Press, 1987): 1–12.

12. See Margaret Kennedy-Dygas, "Historical Perspectives on the 'Science' of Teaching Singing, Part I: Understanding the Anatomy and Function of the Voice (Second through Nineteenth Centuries)," *Journal of Singing* 56, no. 2 (1999): 19–24, and "Historical Perspectives on the 'Science' of Teaching Singing, Part III: Manual Garcia Jr. (1805–1906) and the Science of Singing in the Nineteenth Century," *Journal of Singing* 56, no. 4 (2000): 23–30.

13. This work was produced in two volumes, in 1847 and 1872, and republished as Manuel Garcia and Donald V. Paschke, *A Complete Treatise on the Art of Singing: Complete and Unabridged* (New York: Da Capo, 1975).

14. William Vennard, *Singing: The Mechanism and the Technic* (New York: Carl Fischer, 1968): 1.

15. Ibid., 19.

16. For further views on the relationship between physiological knowledge and vocal pedagogy, see James Q. Davies's fascinating discussion of the relationship between "competing styles of vocal and pianistic presentation, standards of training, and notions of health" (*Romantic Anatomies of Performance* [Berkeley: University of California Press, 2014], 5).

17. And, to start with, because the functions of the larynx change with death, there are tremendous challenges involved in researching it. For an early experiment in keeping the larynx perfused with blood and hence keeping it "alive" in order to learn more about its behavior, see Gerald Berke, Abie H. Mendelsohn, Nelson Scott Howard, and Zhaoyan Zhang, "Neuromuscular Induced Phonation in a Human *Ex Vivo* Perfused Larynx Preparation," *The Journal of the Acoustical Society of America*, 133, no. 2 (2013): EL114–17.

18. Joe Wolfe and Emery Schubert posit that the production of a "fixed," "categorical pitch [that] varies the loudness independently of pitch during single notes"—musical parameters that indeed characterize Western music—is not a natural feature of the voice, given its physiology. Because these sonic characters are "relatively easily produced on simple flute, reed, and string instrument[s]," they posit, vocally extended and fixed pitches were sonic "by-products" gleaned from and imitative of instruments with such sonic characteristics ("Did Non-Vocal Instrument Characteristics Influence Modern Singing?," *Musica Humana* 2, no. 2 [2010]: 121).

19. Quoted in Frances Dyson, *Sounding New Media: Immersion and Embodiment in the Arts and Culture* (Berkeley: University of California Press, 2009), 74.

20. When I speak about this experience in a talk, somebody inevitably comes up to me afterward and shares a similar experience and feeling about it.

21. Philosopher Casey O'Callaghan also posits that sounds are events (*Sound: A Philosophical Theory* [New York: Oxford University Press, 2007]).

22. It is crucial to remember that each vocal activity is not fleeting but physically shapes the vocal apparatus. Vocal activity thus forms the voice and amplifies those vocal sonic possibilities. See, for example, Nina Sun Eidsheim, "Marian Anderson and 'Sonic Blackness' in American Opera," *American Quarterly* 63, no. 3 (2011): 641–71, and "Voice as Action: Towards a Model for Analyzing the Dynamic Construction of Racialized Voice," *Current Musicology* 93, no. 1 (2012), 7–31. Davies has discussed the related question of how music acts in "the cultivation of bodies" in *Romantic Anatomies of Performance*, 2.

23. Michel Foucault, *Discipline and Punish: The Birth of the Prison*, trans. Alan Sheridan (New York: Random House, 1995), 201.

24. For examples of the exercise and instructions discussed in this section, see the video examples on the accompanying website at https://www.youtube.com/watch?v=mKyx4kUa4No&feature=youtu.be (accessed June 1, 2015).

25. With some variation in personal preference, this list moves from basic to advanced: from general pitch, note duration, and syllable to the addition of accents, phrasing, nuanced syllable pronunciation, and timbre. Most singing manuals aimed at voice teachers or students address these parameters, whether directly or indirectly. For example, Oren Brown visualizes them within a twelve-piece pie illustrating the "areas of study that lead to optimum development of singing," which also include posture, breathing, and interpretation (*Discover Your Voice: How to Develop Healthy Voice Habits* [San Diego, CA: Singular, 1996], 118).

26. In the pedagogical scene, there is also a complex of assumptions about the relation between a given pitch and its timbral delivery, or the mediation of that pitch and its expected delivery. In the vocal studio, the piano acts largely as an intermediary between the a priori sound and the student's delivery of it. In other words, the teacher plays a pitch, a run, or a phrase on the piano, and the student delivers it, but the piano's presence and its mediation of the sound have been erased in the process.

27. I am sympathetic to Marcus Boon's argument (*In Praise of Copying* [Cambridge, MA: Harvard University Press, 2010]) that it is the copy that, by definition, creates the original, and that the original is therefore indebted to the copy (which, through its existence, has given the original its status as original) and is in no way superior to it. However, in the value system of the musical community in question, that related to classical music, there is a clear exaltation of the original—as, say, a particular timbre fulfills the notion of classical vocal timbre—and my discussion specifically addresses a scenario in which bodies re-create this timbral ideal.

28. As a side note, even in simply observing the contemporary popular-cultural landscape, we can see that large sectors of the American music and entertainment industry are built around the idea of perfecting imitation of the vocal sounds of particular singers or genres. For example, programs like *American Idol* trade in the currency of a few particular vocal aesthetics, and their narratives are often built around aspiration toward such perfection or its explicit rejection, reactions that may be seen as abject (see Katherine Meizel, *Idolized: Music, Media, and Identity in* American Idol [Blooming-

ton: Indiana University Press, 2011]; Matt Stahl, "A Moment Like This: *American Idol* and Narratives of Meritocracy," in *Bad Music: The Music We Love to Hate*, ed. Chris Washburne and Maiken Derno [New York: Routledge, 2004], 212–32) or as part of a narrative arc built on the rupture between appearance (and associated judgments about the performer) and voice. The UK's *Got Talent* contestants Susan Boyle (season three) and Jonathan and Charlotte (season six) exemplify this phenomenon.

29. Annette Schlichter, "Un/Voicing the Self: Vocal Pedagogy and the Materialization of the Sounding Subject," *Postmodern Culture*, forthcoming.

30. Kristin Linklater, *Freeing the Natural Voice: Imagery and Art in the Practice of Voice and Language*, rev. and expanded ed. (London: Drama, 2006), 7.

31. Ibid., 8. However, as Schlichter observes, "Linklater's voice assemblage is structured through a Romantic ideal of an innocent nature versus a corrupted culture, which entails a whole range of problematic binaries, such as the hierarchies of the 'truth' of a pre-modern Shakespearean past versus a corrupting modernity, intellect versus feeling, live voice versus 'dead' writing, and the body versus technology" ("Un/Voicing the Self"). It is also interesting to note that practice to achieve the "free" voice is dependent on a voice guru.

32. Schlichter, "Un/Voicing the Self." Christina Shewell (*Voice Work: Art and Science in Changing Voices* [Hoboken, NJ: Wiley, 2009]) has used this term to describe clinical or therapeutic vocal training. It is also used to describe the commercial work that vocalists can be hired to carry out, but Schlichter uses it as a specific theoretical term that addresses the inexplicit work required of the voice, such as acting out power dynamics.

33. Schlichter, "Un/Voicing the Self."

34. In a 2011 interview, Linklater reflects on her use of the term "natural voice": "I suppose I am using that controversial word in an oversimplified way, in what has become a cliché, since, of course, we are all formed by both nature and nurture. Is there such a thing as 'natural'? Yet I am interested in the voice as it develops and comes through the child's body, first of all, before it is modified by the environment, by family, culture, or education. So by 'natural' I mean a voice that is in direct connection with emotional impulses. The voice of a baby, in this sense, is 'natural,' and an infant's voice continues to be natural until language comes in, modifies it, shapes it, and very often inhibits the fullness of what that natural voice can communicate" (Dawne McCance, "Crossings: An Interview with Kristin Linklater," *Mosaic* 44, no. 1 [2011]: 21–22).

35. Katherine Bergeron's examination of the formation of *la mélodie française* yields insights on the cross-fertilization between language, education, science, and song that ultimately serves the idea of the nation-state. For example, Ferdinand Brunot's idea of an *Audiothèque*, a large audio collection capturing French speech, containing the "aural heritage of the French people, a *patrimoine sonore*," deeply affected institutionalized phonetic and national pronunciation targets through primary education and thus also took root in the French ear, and was at the root of elocutionary targets for the *mélodie* as well (Katherine Bergeron, *Voice Lessons: French Mélodie in the Belle* Epoque [Oxford: Oxford University Press, 2010], 113). Furthermore, ideology can also be sonified through the appearance of a natural French language, which in fact was built on a few selected

voices and vocalizers before these sounds and articulations were depersonalized through the *Audiothèque* and generalized by mass dissemination. Another powerful example of lips and tongues being charged with a nation-building project may be found in Ana María Ochoa Gautier's work on the connection between spelling and pronunciation in nineteenth-century Colombia. Ochoa outlines Colombia's fascinating eighteenth- and nineteenth-century nation-building project and postcolonial self-definition, in which science (botany) and aesthetics (music) were cross-fertilized, and in which the shaping of vocal sounds played a major role. Specifically, Ochoa interrogates the zoopolitical notion "that the use of the voice to define the boundaries between the human and the non-human generate[s] contested understandings of difference that in turn yielded different notions of the boundaries between music and sound" ("The Zoopolitics of the Voice in Colombia in the Nineteenth Century," the Eleventh Annual Robert Stevenson Lecture at the University of California, Los Angeles, May 2, 2013). In summary, any voice culture, whether in speech or music, is the result of the voice being put to work.

36. Dominic Pettman, "Pavlov's Podcast: The Acousmatic Voice in the Age of MP3s," *differences* 22, no. 2–3 (2011): 158.

37. Mladen Dolar, *A Voice and Nothing More* (Cambridge, MA: MIT Press, 2006), 161.

38. Ibid., 30.

39. For examples of work that discusses this issue, see Anne Carson, "The Gender of Sound," in *Glass, Irony, and God* (New York: New Directions, 1995), 119–42; Cavarero, *For More Than One Voice*; Michelle Duncan, "The Operatic Scandal of the Singing Body: Voice, Presence, Performativity," *Cambridge Opera Journal* 16, no. 3 (2004): 283–306; Nina Sun Eidsheim, "Voice as a Technology of Selfhood: Towards an Analysis of Racialized Timbre and Vocal Performance" (PhD diss., University of California, San Diego, 2008); Katherine Meizel, "A Powerful Voice: Investigating Vocality and Identity," *Voice and Speech Review* 7, no. 1 (2011): 267–74.

40. Catherine Fitzmaurice, "Breathing Is Meaning," in *The Vocal Vision: Views on Voice by 24 Leading Teachers, Coaches and Directors*, ed. Barbara Acker and Marian E. Hampton (New York: Applause, 1997), 247.

41. Because the exercises and scenes I create for my students are far removed from conventional singing situations, this lack of judgment is usually not too challenging to achieve.

42. Michel Chion, *Audio-Vision: Sound on Screen*, trans. Claudia Gorbman (New York: Columbia University Press, 1994), 25, 28. Chion is indebted to Pierre Schaeffer, but Chion's articulation is most relevant for this book. His use of the term "reduced listening"—"a listening mode that focuses on the traits of the sound itself, independent of its cause and its meaning" (ibid., 29)—is also adapted from Schaeffer. Reduced listening assumes that the sound is so complex that it cannot be captured fully in one listening, or in casual listening. Thus repeated listening, made possible with sound recording, is a prerequisite for such reduced listening (ibid.).

43. See, for example, George Thomas Ealy, "Of Ear Trumpets and a Resonance Plate: Early Hearing Aids and Beethoven's Hearing Perception," *19th-Century Music* 17, no. 3

(Spring 1994): 262–73; Brian F. McCabe, "Beethoven's Deafness," *Annals: Official Journal of the American Broncho-Esophagological Association* 67, no. 1 (1958): 192–220; Kenneth M. Stevens and William G. Hemenway, "Beethoven's Deafness," *Journal of the American Medical Association* 213, no. 3 (1970): 434–37.

44. Steven Connor, "Edison's Teeth: Touching Hearing," in *Hearing Cultures: Essays on Sound, Listening and Modernity*, ed. Veit Erlmann (Oxford: Berg, 2004), 153–72. See the introductory chapter, especially note 35's references to the spectrum of sensory experience.

45. Allan L. Benson, "Edison's Dream of New Music," *Cosmopolitan Magazine* LIV (December 1912–May 1913): 798.

46. As paraphrased by Charles Hirschkind, *The Ethical Soundscape: Cassette Sermons and Islamic Counterpublics* (New York: Columbia University Press, 2006), 78.

47. Ibid., 10, see also 32–39.

48. Ibid., 74.

49. Ibid., 76.

50. Ibid., 76.

51. Ibid., 70.

52. Ibid., 84. While Hirschkind deals with a particular cultural and theological context that is different from the context I discuss, in addition to reading sermon listening in the context of the listening practice taught by the Quran, he draws on a theoretical framework that is unconnected to the tradition he studies. In addition to enumerating particular articulations of culture, theology, and media, the general point Hirschkind makes through the listening practice, specific to his case studies, is that there seems to be a relationship between corporeal practice and the desired transformation or maintenance of one's state of mind (or soul or spirit). He derived this general point from Jousse, whose understanding of orality was derived primarily from the study of Christian and Hebrew texts and traditions, and I believe it is also relevant when considering secular corporeal practices.

53. Even in secular listening practices there is an intuitive connection sensed between matter, form, and vocal sound. We see this in situations in which the culturally contingent expected connections and correlations are disrupted—for example, in the perceived disruption of class, race, and gender and their expected sonic pairs. See, for example, Eidsheim, "Marian Anderson and 'Sonic Blackness' in American Opera" and "Voice as Action"; Pettman, "Pavlov's Podcast," 163.

54. There are precedents for this perspective. Others have similarly argued that listening is multisensory and that meaning formation is derived from particular circumstances, informed by broader cultural, historical, and social circumstances. Indeed, this means that that which we have the capacity to hear depends on the circumstantial formation of the sensorium as well as our interpretation of what we hear. See, for example, Judith O. Becker, *Deep Listeners: Music, Emotion, and Trancing* (Bloomington: Indiana University Press, 2004); Tia DeNora, *Music in Everyday Life* (Cambridge: Cambridge University Press, 2000); Dillon, *The Sense of Sound*; Cornelia Fales, "The Paradox of Timbre," *Ethnomusicology* 46, no. 1 (2002): 56–95; Julian Henriques, "The Vibrations of Affect and Their Propagation on a Night out on Kingston's Dancehall Scene," *Body*

and Society 16, no. 1 (2010): 57–89; Hirschkind, *The Ethical Soundscape*; Pauline Oliveros, *Deep Listening: A Composer's Sound Practice* (New York: iUniverse, 2005).

55. Christopher Small, *Musicking: The Meanings of Performing and Listening* (Middletown, CT: Wesleyan University Press, 1998).

56. Brian Massumi, "Too-Blue: Colour-Patch for an Expanded Empiricism," *Cultural Studies* 14, no. 2 (April 2000): 197.

57. Christopher Small, *Musicking: The Meanings of Performing and Listening* (Middletown, CT: Wesleyan University Press, 1998): 2.

58. Benjamin Piekut, *Experimentalism Otherwise: The New York Avant-Garde and Its Limits* (Berkeley: University of California Press, 2011), 159.

Chapter 5. Music as a Vibrational Practice

1. In other words, musical acoustics is the "meeting place of music, vibration physics, auditory science, and craftsmanship" (Arthur H. Benade, *Fundamentals of Musical Acoustics* [New York: Dover, 1990], 3).

2. Terminology such as *the same* rests on assumptions that something already exists and is knowable. While I propose a contrary position throughout this book, I use this language here simply to convey the concept and the position arising from it. I address this problematic in the following paragraph.

3. While I do not throw my hands up when it comes to what is understood as destruction or harm through sound or music, I do believe that the meaning that is derived from, and the effect that is ascribed to, any vibrational node cannot exist prior to a given articulation, and thus a node may be experienced as negative and destructive or positive and restorative.

4. Michel Poizat, *The Angel's Cry: Beyond the Pleasure Principle in Opera*, trans. Arthur Denner (Ithaca, NY: Cornell University Press, 1992), 100.

5. Ibid.

6. Ibid., 101. In psychoanalytical terms, Poizat understands the cry as follows: "A trace of this first satisfaction will remain in the child's psyche, associated with a trace of all the elements—whether feeding, physical contact, or the stimulation of sound, alone or in combination—that bring about the discharge of inner tension. Thereafter the child will have a representation of this object—or group of objects—of its first jouissance" (ibid., 100).

7. Julian Henriques, *Sonic Bodies: Reggae Sound Systems, Performance Techniques, and Ways of Knowing* (New York: Continuum, 2011).

8. Ibid., 242.

9. Ibid.

10. Ibid., 243.

11. Emma Dillon, *The Sense of Sound: Musical Meaning in France, 1260–1330* (Oxford: Oxford University Press, 2012); Shane Butler, *The Hand of Cicero* (London: Routledge, 2002), *The Matter of the Page: Essays in Search of Ancient and Medieval Authors* (Madison: University of Wisconsin Press, 2011), and *The Ancient Phonograph* (New York: Zone Books, 2015).

12. Phil Ford, *Dig: Sound and Music in Hip Culture* (New York: Oxford University Press, 2013), 185, see also 204–10.

13. Poizat, *The Angel's Cry*.

14. Ibid., 39.

15. Ibid., 40. Drawing on Alain Didier-Weill, Poizat sees the blue note as "not only a prelude to the jouissance that this note itself will provide"; like the operatic voice, the blue note "also brings us the promise of jouissance" (ibid., 39).

16. Ibid., 1, 4.

17. David Schwarz, "Listening Subjects: Semiotics, Psychoanalysis, and the Music of John Adams and Steve Reich," *Perspectives of New Music* 31, no. 2 (1993): 24.

18. Ibid., 25, 26.

19. Poizat, *The Angel's Cry*, 48. As the translator, Arthur Denner, explains, Poizat chose to translate the French term *hors-sens* as "trans-sensical" in a way that communicates that it is "beyond-sense": "'Trans-sexual,' 'trans-verbal,' and 'trans-sensical' thus refers [sic] to what stands at the peripheries or in the margins of sexual difference, of words, or of meaning" (Arthur Denner, "Translator's Note," in Michel Poizat, *The Angel's Cry: Beyond the Pleasure Principle in Opera*, trans. Arthur Denner [Ithaca, NY: Cornell University Press, 1992], xiii).

20. Poizat, *The Angel's Cry*, 103. While arrived at through a different framework (voice as intermaterial vibration), my position, described in chapter 4, regarding voice as action resonates with Poizat's. Drawing on the work of Gérard Pommier, Poizat proposes that the child is "dispossessed" of the cry when the other inscribes it within the signifying order—by giving it meaning and acting on that meaning through "com[ing] to alleviate the child's displeasure by feeding it, changing its diapers, or singing it a song" (ibid., 102).

21. Ibid., 104.

22. Ibid.

23. Ibid., 3, 102, 104. What for Poizat is jouissance is, for Schwarz, a presymbolic psychological event and, moving to feminist psychoanalytical theory, is for Renée Cox Lorraine ("Recovering Jouissance: Feminist Aesthetics and Music," in *Women & Music: A History*, ed. Karin Pendle [Bloomington: Indiana University Press, 2001], 3–18) music as "*écriture feminine*," translated to "feminine practice of writing" (Hélène Cixous, "The Laugh of the Medusa," trans. Keith Cohen and Paula Cohen, *Signs* 1, no. 4 (1976): 883). Many debates about the voice within psychoanalytical theory could be cited, but I limit my references to two. First, Rosi Braidotti connects materiality with sound thusly: "the sheer materiality of the human body and its fleshy contents (lungs, nerves, brains, intestines, etc.) are as many sound-making, acoustic chambers" (*Metamorphoses: Towards a Materialist Theory of Becoming* [Cambridge, MA: Polity, 2002], 157). Second, drawing on Braidotti, Milla Tiainen outlines the material-sonic relationship to issues related to sexual differentiation and singing. For Tiainen, the presymbolic and symbolic dynamics may be understood through studying "corporeal materiality in music." That is, "music's bodily aspects are seen as highlighting the major distinction between 'before the Symbolic' and 'after the Symbolic,'" which, she explains, "according to both [Jacques] Lacan and [Julia] Kristeva lies at the core of the subject." Moreover, she continues, "it is this

transition from the pre-symbolic domain to the regulated sphere of language and culture that allegedly creates a crack or a crisis in human (psychic) existence. It is in this way that psychoanalytic approaches, linking music as they do to the longed for, pre-symbolic origins of subjectivity, may be claimed to give its materiality at least a touch of essentiality" ("Corporeal Voices, Sexual Differentiations: New Materialist Perspectives on Music, Singing and Subjectivity," *Thamyris/Intersecting* 18, no. 1 [2008]: 165, note 13). Tiainen's project seeks to recuperate the materiality of the body without essentializing it.

24. Tia DeNora, "Health and Music in Everyday Life—a Theory of Practice," *Psyke and Logos* 28, no. 1 (2007): 271–87; "Interlude 54: Two or More Forms of Music," in *International Handbook of Research in Arts Education*, ed. Liora Bresler (Dordrecht, the Netherlands: Springer, 2007), 799–802; *Music in Everyday Life* (Cambridge: Cambridge University Press, 2000).

25. DeNora, "Interlude 54," 801.

26. Gary Tomlinson, "Evolutionary Studies in the Humanities: The Case of Music," *Critical Inquiry* 39, no. 4 (2013): 647.

27. Ibid., 649.

28. Ibid., 673.

29. John Shepherd's and Peter Wicke's theory of the sonic saddle (a concept they draw from Viktor Zuckerkandl) also describes how the sound of music acts as a structure of feeling. For more on the sonic saddle, see note 96. Considering the 2000 production of *West Side Story* with deaf and hearing cast members by MacMurray College's English and Drama Department, Raymond Knapp reflects on "something graspable on both sides of the hearing divide." In describing the power of musical theater, the cliché of community members' voices joining together is often evoked, but Knapp observes: "In *moving* together, people can know they *belong* together, whatever else divides them, and music has a unique capacity to govern and regulate shared structures of movement. When deaf and hearing attempt to do this together, it may be difficult, but the difficulty is part of the journey, perhaps its most important element, and it is therefore essential that it be manifest in the themes of the musical being performed (as was certainly the case with *West Side Story*)" ("'Waitin' for the Light to Shine': Musicals and Disability," in *The Oxford Handbook of Music and Disability Studies*, ed. Blake Howe, Stephanie Jensen-Moulton, Neil Lerner, and Joseph Straus [New York: Oxford University Press, forthcoming]).

30. See John F. Szwed, *Space Is the Place: The Lives and Times of Sun Ra* (New York: Da Capo, 1998), 121. For an account of the belief of Sun Ra and other 1960s and 1970s musicians in vibrations' vital role in the well-being of the individual and social body, see Gayle Wald, "Soul Vibrations: Black Music and Black Freedom in Sound and Space," *American Quarterly* 63, no. 3 (2011): 673–96.

31. For more on the history of organology, see, for example, Laurence Libin and Renato Meucci, "Organology," *The Grove Dictionary of Musical Instruments*, second edition (New York: Oxford University Press, 2014), 3:753–55. For the recent work on critical organology, see John Tresch and Emily I. Dolan, "Toward a New Organology: Instruments of Music and Science," OSIRIS 28 (2013): 278–98.

32. Incidentally, organology developed at around the same time as Francis Bacon and Galileo Galilei were publishing. Substantial works (including some non-Western) on

instruments were published in the early seventeenth century (Michael Praetorius, *Syntagma Musicum II, De Organographia* [Wolfenbüttel, Germany: Elias Holwein, 1619]). Technical discussions regarding instruments were included in encyclopedic works by, for example, Marin Mersenne (*Harmonie Universelle: The Books on Instruments*, trans. Roger E. Chapman [the Hague, the Netherlands: Martinus Nijhoff, 1957]) and Athanasius Kircher's 1661 *Organum Mathematicum* (see John Edward Fletcher, *A Study of the Life and Works of Athanasius Kircher, "Germanus Incredibilis": With a Selection of His Unpublished Correspondence and an Annotated Translation of His Autobiography* [Leiden; Boston: Brill, 2011]).

However, only after the development of large and permanent instrument collections in Europe and the United States in the nineteenth century did the discipline come to maturity. Organologists, many of whom were also museum curators, could consider comprehensive classification only after they had ready access to such large repositories of instruments. The information in this note is drawn from Libin and Meucci, "Organology."

33. Samuel Brannon, "The Music of the Page: Music Theory, Visual Culture, and Book Design during the Renaissance," Paper presented at the annual meeting of the South Central Graduate Music Consortium, Duke University, September 28, 2013.

34. Galileo Galilei, *Dialogues Concerning Two New Sciences*, trans. Henry Crew and Alfonso de Salvio, with an Introduction by Antonio Favaro (New York: Macmillan, 1914), accessed April 28, 2015, http://oll.libertyfund.org/titles/753.

35. Michael Cyril William Hunter, ed., *Robert Boyle: By Himself and His Friends* (London: Pickering and Chatto, 1994).

36. Bonnie Gordon, "Galileo's Finger and Early Modern Critical Organology" (paper presented at the annual meeting of the American Musicological Society, Pittsburgh, PA, November 7, 2013).

37. Ibid.

38. This succinct formulation can be found at "Bill Brown," accessed April 28, 2015, http://english.uchicago.edu/faculty/brown; also see Bill Brown, "Thing Theory," *Critical Inquiry* 28, no. 1 (2001): 1–22.

39. Shelley Trower, *Senses of Vibration: A History of the Pleasure and Pain of Sound* (New York: Continuum, 2012), 7.

40. Which dimensions of voice concern bodily organs and musical instruments as extensions of the body is a discussion that deserves fuller attention than the arc of this chapter allows. I treat this topic in an article in progress, tentatively titled "Towards an Organology of Musical Matter."

41. Emily Dolan, "Perspectives on Critical Organology," *Newsletter of the American Musical Instrument Society* 43, no. 1 (2014): 14. Dolan adapts the concept of the "boundary object" from Susan Leigh Star and James Greisinger, both of whom, in Dolan's words, "used the term to describe the ways in which the specimens, field notes, and other objects were used by the various social groups that worked in and used Berkeley's Museum of Vertebrate Zoology" (ibid.).

42. Susan Leigh Star and James R. Griesemer, "Institutional Ecology, 'Translations' and Boundary Objects: Amateurs and Professionals in Berkeley's Museum of Vertebrate Zoology, 1907–39," *Social Studies of Science* 19, no. 3 (1989): 408.

43. Trevor Pinch and Frank Trocco, *Analog Days: The Invention and Impact of the Moog Synthesizer* (Cambridge, MA: Harvard University Press, 2002), 349.

44. While the recent interest in instruments can be interpreted as part of a broader humanistic material turn, it is also, of course, a direct continuation of musicology's many assumptions and values. For example, there was a critical organology roundtable titled "Musical Instruments and Cognitions" at the annual meeting of the American Musicological Society in Pittsburgh, PA, November 7–10, 2013.

45. For further reading on these points from outside the fields of engineering and acoustics, see Luciano Chessa, *Luigi Russolo, Futurist: Noise, Visual Arts, and the Occult* (Berkeley: University of California Press, 2012); Steve Goodman, *Sonic Warfare: Sound, Affect, and the Ecology of Fear* (Cambridge, MA: MIT Press, 2010); Douglas Kahn, *Noise, Water, Meat: A History of Sound in the Arts* (Cambridge, MA: MIT Press, 1999); James Gordon Kennaway, *Bad Vibrations: The History of the Idea of Music as Cause of Disease* (Farnham, UK: Ashgate, 2012); Trower, *Senses of Vibration*; Wald, "Soul Vibrations."

46. I want to stress again my belief that investigations based on the premise that music is a social and cultural practice that shares the medium of sound (and its associated technologies and practices) is incredibly useful and important. I also want to stress that I believe knowledge about music is not complete without an inquiry into the cultural, historical, critical, and economic practices within which music is imbricated. In fact, all I want to communicate here is that what we call music operates within multiple sensorial and material realms. Finally, I propose that the divisions represented in figure 5.1 prevent us from acquiring some of that knowledge.

47. Douglas Kahn, *Earth Sound Earth Signal: Energies and Earth Magnitude in the Arts* (Berkeley: University of California Press, 2013), 6.

48. As Kahn notes, Thoreau "heard the music of nature on what he called the *telegraph harp*, a piece of the latest technology acting in an ancient manner"—telegraph lines vibrating in the wind (ibid.).

49. Ibid., 14. Kahn continues, "this was before anyone knew what an antenna was or, for that matter, what electromagnetic radio waves were" (ibid.).

50. Ibid., 6.

51. Singiresu S. Rao, *Mechanical Vibrations*, 5th ed. (Upper Saddle River, NJ: Prentice Hall, 2011). Rao has also coedited a multivolume encyclopedia: Simon Braun, D. J. Ewins, and Singiresu S. Rao, eds., *Encyclopedia of Vibration*, 3 vols. (London: Academic, 2002). For a very useful article on the origins of vibration theory, see Andrew D. Dimarogonas, "The Origins of Vibration Theory," *Journal of Sound and Vibration* 140, no. 2 (1990): 181–89.

52. Singiresu S. Rao, *Vibration of Continuous Systems* (Hoboken, NJ: John Wiley), 2007, 5. A five-hole bone flute found at Hohle Fels in southern Germany indicates that the minimum age of musical artifacts is about 40,000 years, which was roughly the time humans first spread out from Africa to Europe (Nicholas J. Conrad, Maria Malina, and Susanne Münzel, "New Flutes Document the Earliest Musical Tradition in Southwestern Germany," *Nature* 460 [August 2009]: 737–40). According to Tomlinson, the flute's sophistication is evidence of "recognizably modern musicking," which suggests extensive prior musical development ("Evolutionary Studies in the Humanities," 661).

53. Rao, *Vibration of Continuous*, 5.

54. Rao, *Mechanical Vibrations*, 5–11.

55. R. Bruce Lindsay, "The Story of Acoustics," *Journal of the Acoustical Society of America* 39, no. 4 (1966): 629.

56. Ibid., 643–44.

57. Rao, *Mechanical Vibrations*, 5–7, and Lindsay, "The Story of Acoustics," 630–33. For bibliographies of primary sources relevant to the history of vibration and acoustics, see Dimarogonas, "The Origins of Vibration Theory"; Frederick V. Hunt, *Electroacoustics: The Analysis of Transduction, and Its Historical Background* (Cambridge, MA: Harvard University Press, 1954); Lindsay, "The Story of Acoustics"; Dayton Clarence Miller, *Anecdotal History of the Science of Sound to the Beginning of the 20th Century* (New York: Macmillan, 1935); Clifford Truesdell, *The Rational Mechanics of Flexible or Elastic Bodies, 1638–1788: Introduction to Leonhardi Euleri Opera Omnia, Vol. X et XI Seriei Secundae*, vol. 11, sec. 2 (Zürich: Orell Füssli, 1960).

58. Rao, *Mechanical Vibrations*, 11. For a historical summary of vibration studies, including a table summarizing notable contributions from 582–507 BC to 1956, see ibid., 1–13.

59. I began by framing this book as built on the knowledge of artists and practitioners of intermaterial vibration, and the knowledge it is possible to develop as a student and teacher of vibrational practice. While my selection was very much guided by artists who have triggered some of my own insights and who I thought would prove useful heuristically, there are, of course, numerous other artists who could have provided equally good case studies, including people who some readers may have expected to find discussed here. Although I didn't know it when I began to select the material for this book, Kahn's *Earth Sound Earth Signal* is a wonderful companion for my work. Its focus is on electromagnetic waves, or what Kahn refers to as "energetic arts" (ibid., 4), and within this framework, Kahn discusses many artists I do not discuss, but which are relevant to consider under the material vibration framework presented here. Although in his efforts to "think energy," Kahn concentrates on, as his book title offers, "earth sounds and earth signals" instead of "human corporeal energies" (ibid., 9) — with the exception of brain energies in the context of Alvin Lucier's work — we share many concerns, including the ways in which the aesthetic field delineates certain energies or vibrations as falling within its purview while excluding others.

60. National Research Council Committee on a Conceptual Framework for New K–12 Science Education Standards, *A Framework for K–12 Science Education: Practices, Crosscutting Concepts, and Core Ideas* (Washington: National Academies Press, 2012), 132.

61. Sybil P. Parker, *Acoustics Source Book* (New York: McGraw-Hill, 1988), 11. http://www.accessscience.com/content.aspx?SearchInputText=physics+of+sound&id=007000.

62. As mechanical waves are propagated through air into the body, they are mechanically transmitted. But as they are transmitted through the eardrum to the inner ear, there is a mechanical change in the air pressure — that is, a fluctuation in the voltage — and there is a chemical change when the brain changes the level of potassium, which delivers electrons.

63. Chandramohan Sujatha, *Vibration and Acoustics: Measurement and Signal Analysis* (New Delhi: Tata McGraw Hill Education, 2010), 366.

64. DeNora, "Interlude 54," 799.

65. Sujatha, *Vibration and Acoustics*, 296.

66. From this perspective, (1) an instrument is any object that vibrates by default, or (2) an object that is created with the intention of facilitating particular vibrations. Attaching piezoelectric phonograph cartridges to objects to create music, David Tudor's *Rainforest* versions *I* (1968), *II* (1968–69), *III* (1972), and *IV* (1973) to *Forest Speech* (1976) exemplify the former category while musical instruments, such as drums or flutes, exemplify the latter. To read about different versions of *Rainforest* and Tudor's evolving understanding of objects and resonance, see John Driscoll and Matt Rogalsky, "Composers inside Electronics: Music after David Tudor," *Leonardo Music Journal* 14 (2004): 25–30.

67. The numbers in this paragraph are derived from Sujatha, *Vibration and Acoustics*, 294.

68. Vibratory impact on the body is regulated in relation to machines and modes of transportation. Guidelines include Standard 2631–1 (1997) of the International Organization for Standardization (ISO), "Mechanical vibration and shock—Evaluation of human exposure to whole-body vibration," and the British Standard 6841 (1987), "Measurement and evaluation of human exposure to Whole-Body Vibration." The American Acoustical Society publishes national standards on acoustics, including related to mechanical vibration and shock, bioacoustics and noise, as well as a longer list of ISO standards, which are used as international voluntary product safety standards. For more information, see the American Acoustical Society's page about standards on its website: http://acousticalsociety.org/standards/.

69. To read more about spatialization and hearing, see, for example, Jens Blauert, *Spatial Hearing: The Psychophysics of Human Sound Localization*, rev. ed. (Cambridge, MA: MIT Press, 1997).

70. For the specific formula for direct sound source versus reflection, see Blauert, *Spatial Hearing*, 349–50 and 377–78. Also see Hans Wallach, Edwin B. Newman, and Mark R. Rosenzweig, "The Precedence Effect in Sound Localization," *American Journal of Psychology* 62, no. 3 (1949): 315–36; Robert A. Wyttenbach and Ronald R. Hoy, "Demonstration of the Precedence Effect in an Insect," *Journal of the Acoustical Society of America* 94, no. 2 (1993): 777–84.

71. Blauert, *Spatial Hearing*, 4.

72. See the table titled "Psychophysical Theories of Spatial Hearing" in ibid., 13.

73. The ability to identify the difference between a sound source's localization and an auditory event depends on a number of factors, including the listener's familiarity with the signal. Additionally, while it has been established that the node of the eardrum is crucial in determining the relationship between auditory event and sound source, it has now also been established that, on the way to the eardrum, the sound signal undergoes "linear distortions that depend in a characteristic way on the position of the sound source" (ibid., 93). This finding complicates the identification process and, within an intermaterial vibrational organological framework, warrants detailed inquiry. See ibid.,

116–37. Blauert further acknowledges that "it is certainly true that the position of the auditory event and the position of the vibrating body that radiates the sound waves (the sound source) frequently coincide. Nonetheless, the conclusion that the position of the sound source is also the intrinsically correct position of the auditory event is, at the very least, problematic. The sound sources and the auditory event are both sensory objects, after all. If their positions differ, it is an ideal question to ask which is false" (ibid., 4).

74. Ibid., 3–4.
75. Ibid., 37.
76. Ibid., 38.
77. Ibid., 3.
78. Ibid., 2. Blauert is referring to the German Standard DIN 1320 (1959).
79. Ibid. Brian Massumi also makes a similar point vis-à-vis politics in *What Animals Teach Us about Politics* (Durham, NC: Duke University Press, 2014).
80. Examples of scholarly interrogations into assumptions about mechanical vibration or waves, the types of signals or auditory events that constitute musical events, and how humans form meaning around them—which, to my mind, may be considered an organological approach—include Joseph Auner, "Weighing, Measuring, Embalming Tonality: How We Became Phonometrographers," in *Tonality 1900–1950: Concept and Practice*, ed. Feliz Wörner, Ullrich Scheideler, and Philip Rupprecht (Stuttgart, Germany: Franz Steiner Verlag, 2012), 25–46; Katherine Bergeron, *Voice Lessons: French Mélodie in the Belle Epoque* (Oxford: Oxford University Press, 2010); Suzanne G. Cusick, "Acoustemology of Detension in the 'Global War on Terror,'" in *Music, Sound and Space: Transformations of Public and Private Experience*, ed. Georgina Born (Cambridge: Cambridge University Press, 2013), 275–91, "On Musical Performances of Gender and Sex," in *Audible Traces: Gender, Identity and Music*," ed. Elaine Barkin and Lydia Hamessley (Los Angeles: Carciofoli Verlagshaus, 1999), 25–48, and "Re-Soundings: Hearing Worlds from the Global War on Terror," paper presented at the John E. Sawyer Seminar "Hearing Modernity" (Harvard University, November 25, 2013); James Q. Davies, *Romantic Anatomies of Performance* (Berkeley: University of California Press, 2014); Veit Erlmann, *Reason and Resonance: A History of Modern Aurality* (New York: Zone, 2010); Goodman, *Sonic Warfare*; Stefan Helmreich, "An Anthropologist Underwater: Immersive Soundscapes, Submarine Cyborgs, and Transductive Ethnography," *American Ethnologist* 34, no. 4 (2007): 621–41, and "Underwater Music: Tuning Composition to the Sounds of Science," in *The Oxford Handbook of Sound Studies*, ed. Trevor Pinch and Karin Bijsterveld (New York: Oxford University Press, 2012), 151–75; Henriques, *Sonic Bodies*; Kahn, *Earth Sound Earth Signal*; Elisabeth Le Guin, *Boccherini's Body: An Essay in Carnal Musicology* (Berkeley: University of California Press, 2006); Jonathan Sterne, *The Audible Past: Cultural Origins of Sound Reproduction* (Durham, NC: Duke University Press, 2003); Tomlinson, "Evolutionary Studies in the Humanities"; Trower, *Senses of Vibration*.
81. In addition, examples of complex periodic signals include stringed instruments and sawtooth voltage input. Sujatha notes that physical phenomena produce complex periodic signals more commonly than simple harmonic signals (*Vibration and Acoustics*, 4).

82. Knowledge about the piezoelectric effect emerged out of 1880s experiments by the brothers Pierre and Jacques Curie. See Lindsay, "The Story of Acoustics," 634, 641; Pierre Curie, *Œuvres de Pierre Curie, publiées par les soins de la Société Française de Physique* (Paris: Gauthier-Villars, 1908).

83. One example in which the connection between solid states, engineering, vibration, and the performing arts is made palpable may be found in the sound artist Tsunoda Toshiya's 2000 installation. In the piece named "Monitor Unit for Solid Vibration," Toshiya worked from the question, "Is it possible to record or recognize a certain spatial situation or condition using sound?" A positive answer was arrived at by placing a number of piezo ceramic sensors, the technology used in "pickup" microphones, on walls.

84. As I stated above, I believe that in developing analytical frameworks, scholars can gain immensely from observing musical practices from an intensely material point of view. That is, while analytical frameworks for music have not featured robust ways of thinking through music analysis beyond naturalized parameters, artists and instrument and technology developers depend on such currency daily. Kahn discusses artists who deal with "*Energies and Earth Magnitude in the Arts*," as his subtitle puts it (*Earth Sound Earth Signal*). He begins with the story of how radio was heard before its invention when a telephone line picked up the earth's natural electromagnetism, and he follows this thread into a series of accounts of artists' harnessing of the wind (Henry David Thoreau), brainwaves (Alvin Lucier), earthquakes (Hugo Benioff), the sonosphere (Pauline Oliveros), and electromagnetism (Joyce Hinterding), to cite just a few examples.

85. Sujatha observes that "typically vibration measurements are made for the following reasons" and presents a long list, which includes reasons such as "determination of the response of a vibration machine/vehicle/structure" and "finding the response of a human being in a vibrating environment, such as in a vehicle, in a building or while operating a machine" (*Vibration and Acoustics*, 1).

86. Ibid., 15.

87. Thus, "The frequencies of interest in any vibration study are the natural frequencies of the system and the excitation frequencies, which can be related to the running speeds and their harmonics" (ibid., 14).

88. Ibid.

89. Ibid.

90. Ibid., 14–15.

91. Ibid., 15.

92. Rao reminds us: "In fact, most human activities, including hearing, seeing, talking, walking, and breathing, also involve oscillatory motion. Hearing involves vibration of the eardrum, seeing is associated with the vibratory motion of light waves, talking requires oscillation of the laryng [sic] (tongue), walking involves oscillatory motion of legs and hands, and breathing is based on the periodic motions of lungs" (*Mechanical Vibrations*, 4). A telling example of the transposition of vibrations can be found in Mara Mills, "Media and Prosthesis: The Vocoder, the Artificial Larynx, and the History of Signal Processing," *Qui Parle* 21, no. 1 (2012): 107–49.

93. Brian Kane, "Musicophobia, or Sound Art and the Demands of Art Theory," *non-*

site.org, Winter 2012/13, no. 8 (2013). Accessed April 26, 2015, http://nonsite.org/article/musicophobia-or-sound-art-and-the-demands-of-art-theory.

94. Ibid.

95. Ibid.

96. Benjamin Piekut, *Experimentalism Otherwise: The New York Avant-Garde and Its Limits* (Berkeley: University of California Press, 2011), 159.

97. More broadly, we can also say that the world is mediated through a given sensory complex.

98. Judith O. Becker, *Deep Listeners: Music, Emotion, and Trancing* (Bloomington: Indiana University Press, 2004); Steven Connor, "Edison's Teeth: Touching Hearing," in *Hearing Cultures: Essays on Sound, Listening, and Modernity*, ed. Veit Erlmann (Oxford: Berg, 2004), 153–72; DeNora, *Music in Everyday Life*; Goodman, *Sonic Warfare*; Tomie Hahn, *Sensational Knowledge: Embodying Culture through Japanese Dance* (Middletown, CT: Wesleyan University Press, 2007); Henriques, *Sonic Bodies*; Charles Hirschkind, *The Ethical Soundscape: Cassette Sermons and Islamic Counterpublics* (New York: Columbia University Press, 2006); Seth Kim-Cohen, *In the Blink of an Ear: Towards a Non-Cochlear Sonic Art* (New York: Continuum, 2009); Matthew Rahaim, *Musicking Bodies: Gesture and Voice in Hindustani Music* (Middletown, CT: Wesleyan University Press, 2012); Michel Serres, *The Five Senses: A Philosophy of Mingled Bodies*, translated by Margaret Sankey and Peter Cowley (New York: Continuum, 2009).

99. From a physiological point of view, hearing takes place when the hair follicles located in the organ of Corti transduce mechanical sound vibrations into neural impulses.

100. This is an expansion of John Shepherd's and Peter Wicke's theory of the sonic saddle, which takes into account people performing and listening within a semiotic model of timbre. The sonic saddle provides a tool for theorizing, in their words, "how each passing auditory and affective moment can reveal sounds in music acting as structure." Furthermore, the sonic saddle provides a way to conceptualize the overlay of timbral dimensions and internal structures, providing a "continually unfolding sound-image derived from the medium and experienced as the material ground and pathway for the investment of meaning" (*Music and Cultural Theory* [Malden, MA: Blackwell Publishers, 1997], 159–60).

101. Plato puts it this way: "Everything that is responsible for creating something out of nothing is a kind of poetry; and so all the creations of every craft and profession are themselves a kind of poetry, and everyone who practices a craft is a poet" (*Symposium*, trans. Alexander Nehamas and Paul Woodruff [Indianapolis, IN: Hackett, 1995]), 205b).

102. Martin Heidegger, *Poetry, Language, Thought*, trans. Albert Hofstadter (New York: Harper and Row, 1971), 70.

103. Quoted in Chessa, *Luigi Russolo, Futurist*, 1.

104. While I find that what Jean-Luc Nancy accomplishes in thinking through sound's resonance and its implications for form and subjectivity is productive, I also find it rather unfortunate that Nancy reverts to what Sterne has termed the "audiovisual litany," a false polar opposition between sound and visuality (*The Audible Past*, 15–19). For a care-

ful reading of Nancy's usage of the French verbs *écouter* ("to hear") and *entendre* ("to listen"), see Brian Kane, "Jean-Luc Nancy and the Listening Subject," *Contemporary Music Review* 31, nos. 5–6 (2012): 439–47. In other words, for sound to be invited forward onto the stage (to use a visual metaphor), visuality is thrown under the bus.

105. This position is likely to be seen as extreme by most readers. However, there are plenty of examples showing how what we very easily could agree constitutes a human being, with attendant responsibilities, has not remained constant throughout history. The differentiation between a free person and a slave is an extreme example. While most of us would say that such an example is based on a position regarding human versus nonhuman with which we do not agree, and for which we have no reference, we can relate to the way in which a mass of people is not met with the same empathy (multiplied by the number of people) as a single person who is in front of us.

106. In my use of the term *zone*, I draw on the composer and trumpeter Wadada Leo Smith. In rehearsals, Smith often describes sections of pieces, textures, modes, feels, phrasings, and more as "zones."

107. Many of the most powerful vocal performances in which I have taken part were events for which no entry tickets were issued. Being together without identification, being present for one another without attachment to names—being nothing and everything—frees the heart to sing and to receive that song.

BIBLIOGRAPHY

Abbate, Carolyn. *In Search of Opera*. Princeton, NJ: Princeton University Press, 2001.
———. "Music—Drastic or Gnostic?" *Critical Inquiry* 30, no. 3 (2004): 505–36.
———. *Unsung Voices: Opera and Musical Narrative in the Nineteenth Century*. Princeton, NJ: Princeton University Press, 1991.
"About the Ig Nobel Prizes." Improbable Research. Accessed March 7, 2015. http://www.improbable.com/ig/.
Absuperman. "Speech Jammer Gun Review (Original)." May 31, 2013. Accessed June 1, 2013. http://www.youtube.com/watch?v=oU9EGeMP5n4.
Adorno, Theodor. *Essays on Music*. Edited by Richard Leppert and translated by Susan H. Gillespie. Berkeley: University of California Press, 2002.
Akers, Matthew, and Jeff Dupre, dirs. *Marina Abramović: The Artist Is Present*. Chicago: Music Box Films, 2012. DVD.
Altman, Rick. *Sound Theory, Sound Practice*. New York: Routledge, 1992.
Amacher, Maryanne. "Maryanne Amacher in Conversation with Frank J. Oteri." April 16, 2004, Kingston, NY, videotaped by Randy Nordschowl, transcribed by Molly Sheridan and Randy Nordschow. Accessed April 21, 2015. https://www.newmusicbox.org/assets/61/interview_amacher.pdf.
Appleton, Ian. *Buildings for the Performing Arts: A Design and Development Guide*. Boston: Butterworth Architecture, 1996.
Aristotle. *The Complete Works of Aristotle*. Edited by Jonathan Barnes, translated by D'Arcy Wentworth Thompson. Princeton, NJ: Princeton University Press, 1984.
Arnheim, Rudolf. *Radio*. London: Faber and Faber, 1936.
"Artbound Special Episode: 'Invisible Cities.'" KCET. January 4, 2015. Accessed March 6, 2015. http://www.kcet.org/arts/artbound/counties/los-angeles/artbound-special-episode-invisible-cities.html.
Augustine. *St. Augustine: Confessions*. Translated by Henry Chadwick. Oxford: Oxford University Press, 1991.
Auner, Joseph. "Weighing, Measuring, Embalming Tonality: How We Became Phonometrographers." In *Tonality 1900–1950: Concept and Practice*, edited by Feliz Wörner,

Ullrich Scheideler, and Philip Rupprecht, 25–46. Stuttgart, Germany: Franz Steiner Verlag.

Austern, Linda Phyllis, and Inna Naroditskaya, eds. *Music of the Sirens*. Bloomington: Indiana University Press, 2006.

Banes, Sally. "The Art of Meredith Monk." *Performing Arts Journal* 3, no. 1 (1978): 3–18.

Banister, Harry. "A Further Note on the Phase Effect in the Localization of Sound." *British Journal of Psychology* 15, no. 1 (1924): 80–81.

Barad, Karen Michelle. *Meeting the Universe Halfway: Quantum Physics and the Entanglement of Matter and Meaning*. Durham, NC: Duke University Press, 2007.

Bard, Perry, and Alejandro Jaramillo. "Boomerang: No Delay." February 8, 2011. Accessed April 1, 2012. http://www.youtube.com/watch?v=ERobX2de5DQ.

Barrett, Estelle, and Barbara Bolt. *Carnal Knowledge: Towards a 'New Materialism' through the Arts*. London; New York: I.B. Tauris, 2013.

Barthes, Roland. "The Grain of the Voice." In *Image, Music, Text*, essays selected and translated by Stephen Heath, 179–89. New York: Hill and Wang, 1977.

Beahrs, Robert O. "Post-Soviet Tuvan Throat-Singing (Xöömei) and the Circulation of Nomadic Sensibility." PhD diss., University of California, Berkeley, 2014.

Beauvoir, Simone de. *The Second Sex*. New York: Vintage, 1974.

Becker, Judith O. *Deep Listeners: Music, Emotion, and Trancing*. Bloomington: Indiana University Press, 2004.

Benade, Arthur H. *Fundamentals of Musical Acoustics*. New York: Dover, 1990.

Bennett, Jane. *Vibrant Matter: A Political Ecology of Things*. Durham, NC: Duke University Press, 2010.

Benson, Allan L. "Edison's Dream of New Music." *Cosmopolitan Magazine* LIV (December 1912–May 1913): 797–800.

Beranek, Leo Leroy. *Acoustic Measurements*. New York: John Wiley, 1949.

———. "Analysis of Sabine and Eyring Equations and Their Application to Concert Hall Audience and Chair Absorption." *Journal of the Acoustical Society of America* 120, no. 3 (2006): 1399–410.

———. *Concert Halls and Opera Houses: Music, Acoustics, and Architecture*. Rev. ed. New York: Springer, 2004.

———. *Music, Acoustics and Architecture*. New York: Wiley, 1962.

———. "Subjective Rank-Ordering and Acoustical Measurements for Fifty-Eight Concert Halls." *Acta Acustica United with Acustica* 89, no. 3 (2003): 494–508.

Bergeron, Katherine. *Voice Lessons: French Mélodie in the Belle Epoque*. Oxford: Oxford University Press, 2010.

Berke, Gerald, Abie H. Mendelsohn, Nelson Scott Howard, and Zhaoyan Zhang. "Neuromuscular Induced Phonation in a Human Ex Vivo Perfused Larynx Preparation." *Journal of the Acoustic Society of America* 133, no. 2 (2013): EL114–17.

Berkeley, George. *Principles of Human Knowledge and Three Dialogues between Hylas and Philonous*. Edited by Roger Woolhouse. London: Penguin, 1988.

Bernard, Jonathan. "Feldman's Painters." In *The New York Schools of Music and Visual Arts: John Cage, Morton Feldman, Edgard Varèse, Willem De Kooning, Jasper Johns, Robert Rauschenberg*, edited by Steven Johnson, 173–215. New York: Routledge, 2002.

"Biography." Meredith Monk. Accessed April 1, 2013. http://www.meredithmonk.org/about/bio.html.
Blauert, Jens. *Spatial Hearing: The Psychophysics of Human Sound Localization*. Rev. ed. Cambridge, MA: MIT Press, 1997.
———. "Untersuchungen zum Richtungshören in der Medianebene bei fixiertem Kopf" [Investigations of directional hearing in the median plane with the head immobilized]. PhD diss., Technische Hochschule, Aachen, 1969.
Blesser, Barry, and Linda-Ruth Salter. "Ancient Acoustic Spaces." In *The Sound Studies Reader*, edited by Jonathan Sterne, 186–96. New York: Routledge, 2012.
Bloechl, Olivia Ashley. *Native American Song at the Frontiers of Early Modern Music*. Cambridge: Cambridge University Press, 2008.
Bois, Yve-Alain. "1955a." In *Art since 1900: Modernism, Antimodernism, Postmodernism*, edited by Hal Foster, Rosalind Krauss, Yve-Alain Bois, and Benjamin H. D. Buchloh, 2:411–13. London: Thames and Hudson, 2004.
Boon, Marcus. *In Praise of Copying*. Cambridge, MA: Harvard University Press, 2010.
Born, Georgina. "For a Relational Musicology: Music and Interdisciplinarity, Beyond the Practice Turn." *Journal of the Royal Music Association* 135, no. 2 (2010): 205–43.
———. Introduction to *Music, Sound and Space: Transformations of Public and Private Experience*, edited by Georgina Born, 1–70. Cambridge: Cambridge University Press, 2013.
Bosma, Hannah. "The Electronic Cry: Voice and Gender in Electroacoustic Music." PhD diss., University of Amsterdam, 2013.
Brackett, David. *Interpreting Popular Music*. Cambridge: Cambridge University Press, 1995.
Braidotti, Rosi. *Metamorphoses: Towards a Materialist Theory of Becoming*. Cambridge, MA: Polity, 2002.
Brakhage, Stan, and Joseph Cornell, dirs. *By Brakhage: An Anthology, Volume One*. Los Angeles: Criterion, 2010. DVD.
Brannon, Samuel. "The Music of the Page: Music Theory, Visual Culture, and Book Design during the Renaissance." Paper presented at the annual meeting of the South Central Graduate Music Consortium. Duke University, September 28, 2013.
Braun, Simon, D. J. Ewins, and Singiresu S. Rao, eds. *Encyclopedia of Vibration*. 3 vols. London: Academic, 2002.
Brekhovskikh, L. M., and Yu. P. Lysanov. *Fundamentals of Ocean Acoustics*. 3rd ed. New York: Springer-Verlag, 2003.
Brown, Bill. "Thing Theory." *Critical Inquiry* 28, no. 1 (2001): 1–22.
Brown, Oren. *Discover Your Voice: How to Develop Healthy Voice Habits*. San Diego, CA: Singular, 1996.
Bulut, Zeynep. "La Voix-Peau: Understanding the Physical, Phenomenal and Imaginary Limits of the Human Voice through Contemporary Music." PhD diss., University of California, San Diego, 2011.
Butler, Shane. *The Ancient Phonograph*. New York: Zone Books, 2015.
———. *The Hand of Cicero*. London: Routledge, 2002.

———. *The Matter of the Page: Essays in Search of Ancient and Medieval Authors.* Madison: University of Wisconsin Press, 2011.

———. "What Was the Voice?" Unpublished manuscript.

Cage, John. *4′33″: For Any Instrument or Combination of Instruments.* New York: Henmar Press, 1993.

———. *Silence: Lectures and Writings.* 1st ed. Middletown, CO: Wesleyan University Press, 1961.

Calame-Griaule, Genevieve. "Voice and the Dogon World." In *Notebooks in Cultural Analysis*, edited by Norman F. Cantor and Nathalia King, 15–60. Durham, NC: Duke University Press, 1986.

Calvino, Italo. *Invisible Cities.* Translated by William Weaver. Orlando, FL: Harcourt Brace, 1972.

Canguilhem, Georges. *On the Normal and the Pathological.* Translated by Carolyn R. Fawcett, and introduction by Michel Foucault. Dordrecht, the Netherlands: D. Reidel, 1978.

Cannella, Gaile Sloan, Michelle Salazar Pérez, and Penny A. Pasque. *Critical Qualitative Inquiry: Foundations and Futures.* Walnut Creek, CA: Left Coast Press, 2015.

Carney, Dana R., Amy J. C. Cuddy, and Andy J. Yap. "Power Posing: Brief Nonverbal Displays Affect Neuroendocrine Levels and Risk Tolerance." *Psychological Science* 21, no. 10 (2012): 1363–68.

Carpenter, Alexander. "Schoenberg's Vienna, Freud's Vienna: Re-Examining the Connections between the Monodrama *Erwartung* and the Early History of Psychoanalysis." *Musical Quarterly* 93, no. 1 (2010): 144–81.

Carson, Anne. "The Gender of Sound." In Anne Carson, *Glass, Irony, and God*, introduction by Guy Davenport, 119–42. New York: New Directions, 1995.

Cavarero, Adriana. *For More Than One Voice: Toward a Philosophy of Vocal Expression.* Translated by Paul A. Kettman. Stanford, CA: Stanford University Press, 2005.

Cerrone, Christopher. "Invisible Cities: Composing an Opera for Headphones." Artbound. October 22, 2013. Accessed March 24, 2014. http://www.kcet.org/arts/artbound/counties/los-angeles/invisible-cities-opera-composer-christopher-cerrone-union-station-music.html.

Chamberlain, David B. "The Fetal Sense: A Classical View." Accessed April 21, 2015. https://birthpsychology.com/free-article/fetal-senses-classical-view.

Chessa, Luciano. *Luigi Russolo, Futurist: Noise, Visual Arts, and the Occult.* Berkeley: University of California Press, 2012.

Chion, Michel. *Audio-Vision: Sound on Screen.* Translated by Claudia Gorbman. New York: Columbia University Press, 1994.

Cixous, Hélène. "The Laugh of the Medusa." Translated by Keith Cohen and Paula Cohen. *Signs* 1, no. 4 (1976): 875–93.

Collins, Billy. TED *Radio Hour.* National Public Radio. June 1, 2012. Accessed April 17, 2015, http://www.npr.org/templates/transcript/transcript.php?storyId=153699514.

Connor, Steven. *Dumbstruck: A Cultural History of Ventriloquism.* Oxford: Oxford University Press, 2000.

———. "Edison's Teeth: Touching Hearing." In *Hearing Cultures: Essays on Sound, Listening and Modernity*, edited by Veit Erlmann, 153–72. Oxford: Berg, 2004.

———. "Michel Serres's Five Senses." 1999. Accessed June 19, 2012. http://www.stevenconnor.com/5senses.htm.

Conard, Nicholas J., Maria Malina, and Susanne Münzel. "New Flutes Document the Earliest Musical Tradition in Southwestern Germany." *Nature* 460 (August 2009): 737–40.

Coole, Diana H., and Samantha Frost. *New Materialisms: Ontology, Agency, and Politics*. Durham, NC; London: Duke University Press, 2010.

Copenhafer, David Tyson, IV. "Invisible Ink: Philosophical and Literary Fictions of Music." PhD diss., University of California, Berkeley, 2004.

Cox Lorraine, Renée. "Recovering Jouissance: Feminist Aesthetics and Music." In *Women and Music: A History*, edited by Karin Pendle, 3–18. Bloomington: Indiana University Press, 2001.

Craig, Daniel G. "An Exploratory Study of Physiological Changes during 'Chills' Induced by Music." *Musicae Scientiae* 9, no. 2 (2005): 273–87.

Cruz, Jon. *Culture on the Margins: The Black Spiritual and the Rise of American Cultural Interpretation*. Princeton, NJ: Princeton University Press, 1999.

Csordas, Thomas J. "Embodiment as a Paradigm for Anthropology." *Ethos* 18, no. 1 (1990): 5–47.

Culler, Jonathan D. *On Deconstruction: Theory and Criticism after Structuralism*. Ithaca, NY: Cornell University Press, 2007.

Cummins, Fred. "Rhythm as an Affordance for the Entrainment of Movement." *Phonetica* 66, nos. 1–2 (2009): 15–28.

Curie, Pierre. *Œuvres de Pierre Curie, publiées par les soins de la Société Française de Physique*. Paris: Gauthier-Villars, 1908.

Cusick, Suzanne G. "Acoustemology of Detension in the 'Global War on Terror.'" In *Music, Sound and Space: Transformations of Public and Private Experience*, edited by Georgina Born, 275–91. Cambridge: Cambridge University Press, 2013.

———. "On Musical Performances of Gender and Sex." In *Audible Traces: Gender, Identity and Music*, edited by Elaine Barkin and Lydia Hamessley, 25–48. Los Angeles: Carciofoli Verlagshaus, 1999.

———. "Re-Soundings: Hearing Worlds from the Global War on Terror." Paper presented at the Sawyer Seminar "Hearing Modernity," Harvard University, November 25, 2013.

Damasio, Antonio R. *Descartes' Error: Emotion, Reason, and the Human Brain*. New York: G. P. Putnam, 1994.

———. *The Feeling of What Happens: Body and Emotion in the Making of Consciousness*. New York: Harcourt Brace, 1999.

Dart, Thurston. *The Interpretation of Music*. London: Hutchinson's University Library, 1954.

Davies, James Q. *Romantic Anatomies of Performance*. Berkeley: University of California Press, 2014.

Davis, Mike. *City of Quartz: Excavating the Future in Los Angeles*. London: Verso, 1990.

Delle Sedie, Enrico. *A Complete Method of Singing: A Theoretical and Practical Treatise on the Art of Singing*. New York: George Schirmer, 1894.

DeNora, Tia. *After Adorno: Rethinking Music Sociology*. Cambridge: Cambridge University Press, 2003.

———. "Health and Music in Everyday Life—a Theory of Practice." *Psyke and Logos* 28, no. 1 (2007): 271–87.

———. "Interlude 54: Two or More Forms of Music." In *International Handbook of Research in Arts Education*, edited by Liora Bresler, 799–802. Dordrecht, the Netherlands: Springer, 2007.

———. *Music in Everyday Life*. Cambridge: Cambridge University Press, 2000.

Derrida, Jacques. *Of Grammatology*. Translated by Gayatri Chakravorty Spivak. Baltimore, MD: Johns Hopkins University Press, 1976.

———. *Positions*. Translated by Alan Bass. Chicago: University of Chicago Press, 1981.

Dillon, Emma. *The Sense of Sound: Musical Meaning in France, 1260–1330*. New York: Oxford University Press, 2012.

Dimarogonas, Andrew D. "The Origins of Vibration Theory." *Journal of Sound and Vibration* 140, no. 2 (1990): 181–89.

Dolan, Emily I. "Perspectives on Critical Organology." *Newsletter of the American Musical Instrument Society* 43, no. 1 (2014): 14–16.

Dolar, Mladen. *A Voice and Nothing More*. Cambridge, MA: MIT Press, 2006.

Driscoll, John, and Matt Rogalsky. "Composers inside Electronics: Music after David Tudor." *Leonardo Music Journal* 14 (December 2004): 25–30.

Duncan, Michelle. "The Operatic Scandal of the Singing Body: Voice, Presence, Performativity." *Cambridge Opera Journal* 16, no. 3 (2004): 283–306.

Dunn, Leslie C., and Nancy A. Jones. *Embodied Voices: Representing Female Vocality in Western Culture*. Cambridge: Cambridge University Press, 1994.

Dyson, Frances. *Sounding New Media: Immersion and Embodiment in the Arts and Culture*. Berkeley: University of California Press, 2009.

Ealy, George Thomas. "Of Ear Trumpets and a Resonance Plate: Early Hearing Aids and Beethoven's Hearing Perception." *19th-Century Music* 17, no. 3 (1994): 262–73.

"Editor's Table." *Chautauquan*, June 1883, 543–44.

Eidsheim, Nina Sun. "*Invisible Cities* at Union Station, Los Angeles 2014." Unpublished manuscript.

———. "Marian Anderson and 'Sonic Blackness' in American Opera," *American Quarterly* 63, no. 3 (2011): 641–71.

———. "Measuring Race: Listening to Vocal Timbre and Vocality in African-American Popular Music." Unpublished manuscript.

———. "Race and Aesthetics of Vocal Timbre." In *Rethinking Difference in Music Scholarship*, edited by Olivia Bloechl, Melanie Lowe, and Jeffrey Kallberg, 338–65. Cambridge: Cambridge University Press, 2015.

———. "Sensing Voice: Materiality and Presence in Singing and Listening." *Senses and Society* 6, no. 2 (2011): 133–55.

———. "Voice as a Technology of Selfhood: Towards an Analysis of Racialized Timbre and Vocal Performance." PhD diss., University of California, San Diego, 2008.

———. "Voice as Action: Towards a Model for Analyzing the Dynamic Construction of Racialized Voice." *Current Musicology* 93, no. 1 (2012), 7–31.

Eidsheim, Nina Sun, and Katherine Meizel, eds. *The Oxford Handbook of Voice Studies*. New York: Oxford University Press, forthcoming.

Eidsheim, Nina Sun, and Annette Schlichter, eds. Special issue on voice and materiality, *Postmodern Culture*, forthcoming.

Einstein, Albert. "Does the Inertia of a Body Depend upon Its Energy Content?" In *The Collected Papers of Albert Einstein*, vol. 2, *The Swiss Years: Writings, 1900–1909*, translated by Anna Beck, 174. Princeton, NJ: Princeton University Press, 1989.

———. *The World as I See It*. Translated by Alan Harris. New York: Philosophical Library, 1949.

Erlmann, Veit. *Reason and Resonance: A History of Modern Aurality*. New York: Zone, 2010.

Esposito, Christina M. "The Effects of Linguistic Experience on the Perception of Phonation." *Journal of Phonetics* 38, no. 2 (2010): 306–16.

Fales, Cornelia. "The Paradox of Timbre." *Ethnomusicology* 46, no. 1 (2002): 56–95.

Feld, Steven. *Sound and Sentiment: Birds, Weeping, Poetics, and Song in Kaluli Expression*. 2nd ed. Philadelphia: University of Pennsylvania Press, 1990.

Feldman, Martha. *The Castrato: Reflections on Natures and Kinds*. Berkeley: University of California Press, 2014.

Feldman, Martha, and Bonnie Gordon, eds. *The Courtesan's Arts: Cross-Cultural Perspectives*. New York: Oxford University Press, 2006.

Fink, Robert. "Unwrapping the Box: Gehry's Walt Disney Concert Hall as Postmodern Space." Paper presented at the annual meeting of the American Musicological Society, Philadelphia, PA, November 12–15, 2009.

Fisher, George, and Judy Lochhead. "Analyzing from the Body." *Theory and Practice* 27 (2002): 37–68.

Fitzmaurice, Catherine. "Breathing Is Meaning." In *The Vocal Vision: Views on Voice by 24 Leading Teachers, Coaches and Directors*, edited by Barbara Acker and Marian E. Hampton, 247–52. New York: Applause, 1997.

Ford, Phil. *Dig: Sound and Music in Hip Culture*. New York: Oxford University Press, 2013.

Forsyth, Michael. *Buildings for Music: The Architect, the Musician, and the Listener from the Seventeenth Century to the Present Day*. Cambridge, MA: MIT Press, 1985.

Foster, Hal, Rosalind Krauss, Yves-Alain Bois, and Benjamin H. D. Buchloh, eds. *Art since 1900: Modernism, Antimodernism, Postmodernism*. Vol. 2. London: Thames and Hudson, 2004.

Foucault, Michel. *Discipline and Punish: The Birth of the Prison*. Translated by Alan Sheridan. New York: Random House, 1995.

Galileo Galilei. *Two New Sciences, Including Centers of Gravity and Force of Percussion*. Translated by Stillman Drake. Madison: University of Wisconsin Press, 1974.

Garcia, Manuel. *A Complete Treatise on the Art of Singing: Complete and Unabridged*, edited by Donald V. Paschke. New York: Da Capo, 1975.

Garcia, Phillip. "How to: Speech Jam / Jammer with Audacity." March 14, 2013. Accessed July 16, 2013. https://www.youtube.com/watch?v=h057bsIIoo4.

Geertz, Clifford. *The Interpretation of Cultures*. New York: Basic, 1973.

Gick, Bryan, and Donald Derrick. "Aero-Tactile Integration in Speech Perception." *Nature* 462, no. 7272 (2009): 502–4.

Gjerdingen, Robert O., and David Perrott. "Scanning the Dial: The Rapid Recognition of Music Genres." *Journal of New Music Research* 37, no. 2 (2008): 93–100.

Goehr, Lydia. *The Imaginary Museum of Musical Works: An Essay in the Philosophy of Music*. Oxford: Clarendon Press of Oxford University Press, 1992.

Goodman, Steve. *Sonic Warfare: Sound, Affect, and the Ecology of Fear*. Cambridge, MA: MIT Press, 2010.

Gordon, Bonnie. "Galileo's Finger and Early Modern Critical Organology." Paper presented at the annual meeting of the American Musicological Society, Pittsburgh, PA, November 7, 2013.

———. *Monteverdi's Unruly Women: The Power of Song in Early Modern Italy*. Cambridge: Cambridge University Press, 2004.

Griffiths, Theresa, dir. *Jackson Pollock: Love and Death on Long Island*. London: BBC, 1999. Accessed April 1, 2013. http://www.bbc.co.uk/programmes/b0074km1.

Guck, Marion A. "Who Counts?" Paper presented at the joint meeting of the American Musicological Society, Society for Ethnomusicology, and Society for Music Theory, New Orleans, LA, November 3, 2012.

Hahn, Tomie. *Sensational Knowledge: Embodying Culture through Japanese Dance*. Middletown, CT: Wesleyan University Press, 2007.

Hammond, Michael. *Performing Architecture: Opera Houses, Theatres and Concert Halls for the Twenty-First Century*. London: Merrell, 2006.

Harada, Kai. "Opera's Dirty Little Secret." *Live Design*. March 1, 2001. Accessed April 1, 2012. http://livedesignonline.com/mag/operas-dirty-little-secret.

Heidegger, Martin. *Poetry, Language, Thought*. Translated by Albert Hofstadter. New York: Harper and Row, 1971.

Helmreich, Stefan. "An Anthropologist Underwater: Immersive Soundscapes, Submarine Cyborgs, and Transductive Ethnography." *American Ethnologist* 34, no. 4 (2007): 621–41.

———. "Underwater Music: Tuning Composition to the Sounds of Science." In *The Oxford Handbook of Sound Studies*, edited by Trevor Pinch and Karin Bijsterveld, 151–75. New York: Oxford University Press, 2012.

Henriques, Julian. *Sonic Bodies: Reggae Sound Systems, Performance Techniques, and Ways of Knowing*. New York: Continuum, 2011.

———. "The Vibrations of Affect and Their Propagation on a Night out on Kingston's Dancehall Scene." *Body and Society* 16, no. 1 (2010): 57–89.

Hepper, Peter G., and B. Sara Shahidullah. "Development of Fetal Hearing." *Archives of Disease in Childhood—Fetal and Neonatal Edition* 71, no. 2 (September 1, 1994): F81–F87.

Hersey, John B. "A Chronicle of Man's Use of Ocean Acoustics." *Oceanus* 20, no. 2 (1977): 8–21.

Hickson, Frances S., and Valerie E. Newton. "Sound Localization: Part I." *Journal of Laryngology and Otology* 95, no. 1 (1981): 29–40.

Hirschkind, Charles. *The Ethical Soundscape: Cassette Sermons and Islamic Counterpublics.* New York: Columbia University Press, 2006.

Holsinger, Bruce W. *Music, Body, and Desire in Medieval Culture: Hildegard of Bingen to Chaucer.* Stanford, CA: Stanford University Press, 2001.

Horkheimer, Max, and Theodor W. Adorno. *Dialectic of Enlightenment: Philosophical Fragments.* Translated by Gunzelin Schmid Noerr. Stanford, CA: Stanford University Press, 2002.

Hower, Noble. "Speech Jammer with the Hower Family." December 28, 2012. Accessed June 1, 2013. http://www.youtube.com/watch?v=X6nypeuC4j4.

Hughes, Robert. "An American Legend in Paris." *Time* 119, no. 5 (February 1982): 76.

Hunt, Frederick V. *Electroacoustics: The Analysis of Transduction, and Its Historical Background.* Cambridge, MA: Harvard University Press, 1954.

Hunter, Michael Cyril William, ed. *Robert Boyle: By Himself and His Friends.* London: Pickering and Chatto, 1994.

Hurd, Michael, and John Borwick. "Concert Halls." In *The Oxford Companion to Music*, edited by Alison Latham, 287–88. New York: Oxford University Press, 2002.

Ihde, Don. *Listening and Voice: A Phenomenology of Sound.* Athens: Ohio University Press, 1976.

———. *Listening and Voice: Phenomenologies of Sound.* 2nd ed. Albany: State University of New York Press, 2007.

Institute for Research and Coordination in Acoustics/Music. "Who Are We?" Accessed April 1, 2014. http://www.ircam.fr/ircam.html?&L=1.

Iverson, Paul, et al. "A Perceptual Interference Account of Acquisition Difficulties for Non-Native Phonemes." *Cognition* 87, no. 1 (2003): B47–B57.

Izdebski, Krzystof. *Emotions in the Human Voice.* San Diego, CA: Plural, 2008.

Izenour, George C., Vern Oliver Knudsen, and Robert B. Newman. *Theater Design.* New York: McGraw-Hill, 1977.

Jaffe, Christopher J. *The Acoustics of Performance Halls: Spaces for Music from Carnegie Hall to the Hollywood Bowl.* New York: W. W. Norton, 2010.

Jaffe, Christopher J., Mark Holden, and Robert Lilkendey. "Consistency in Acoustic Design When Building Symphonic Performance Halls." Accessed June 20, 2014. http://www.siebeinacoustic.com/publications/papers/1999%20-%20Consistency%20in%20Acoustic%20Design%20When%20Building.pdf.

Jakobson, Roman. "Why 'Mama' and 'Papa'?" In Roman Jakobson, *Studies on Child Language and Aphasia*, 21–29. The Hague, the Netherlands: Mouton, 1971.

jim477. "Rhett and Link speech jammer (Garageband)." April 28, 2015, https://www.youtube.com/watch?v=mXFsjqlo8rc.

Johnstone-Douglas, William. "The Teaching of Jean de Reszke." In *Historical Vocal Pedagogy Classics*, edited by Berton Coffin, 104–11. Lanham, MD: Scarecrow, 1989.

Jones, Amelia. "Holy Body: Erotic Ethics in Ron Athey and Juliana Snapper's Judas Cradle," *TDR* 50, no. 1 (2006): 159–69.

———. *Performing the Body/Performing the Text.* London: Routledge, 1999.

Jowitt, Deborah. *Meredith Monk*. Baltimore, MD: Johns Hopkins University Press, 1997.

———. "Monk and King: The Sixties Kids." In *Reinventing Dance in the 1960s: Everything Was Possible*, edited by Sally Banes, Andrea Harris, and Mikhail Baryshnikov, 113–36. Madison: University of Wisconsin Press, 2003.

———. *Time and the Dancing Image*. Berkeley: University of California Press, 1989.

Jul Snapper. "Juliana Snapper_work.excerpts." April 21, 2015, https://vimeo.com/81368087.

Kafka, Franz. *The Complete Stories*. Translated by Nahum Norbert Glatzer. New York: Schocken, 1995.

Kahn, Douglas. *Earth Sound Earth Signal: Energies and Earth Magnitude in the Arts*. Berkeley: University of California Press, 2013.

———. *Noise, Water, Meat: A History of Sound in the Arts*. Cambridge, MA: MIT Press, 1999.

Kane, Brian. "Jean-Luc Nancy and the Listening Subject." *Contemporary Music Review* 31, nos. 5–6 (2012): 439–47.

———. "Musicophobia, or Sound Art and the Demands of Art Theory." *Nonsite.org*, no. 8 (Winter 2012/2013). Accessed April 26, 2015, http://nonsite.org/article/musicophobia-or-sound-art-and-the-demands-of-art-theory.

———. *Sound Unseen: Acousmatic Sound in Theory and Practice*. New York: Oxford University Press, 2014.

Kaprow, Allan. *Assemblage, Environments, Happenings*. New York: Harry N. Abrams, 1966.

Kelman, Ari V. "Rethinking the Soundscape: A Critical Genealogy of a Key Term in Sound Studies." *Senses and Society* 5, no. 2 (2010): 212–34.

Kennaway, James Gordon. *Bad Vibrations: The History of the Idea of Music as Cause of Disease*. Farnham, UK: Ashgate, 2012.

Kennedy-Dygas, Margaret. "Historical Perspectives on the 'Science' of Teaching Singing, Part I: Understanding the Anatomy and Function of the Voice (Second through Nineteenth Centuries)." *Journal of Singing* 56, no. 2 (1999): 19–24.

———. "Historical Perspectives on the 'Science' of Teaching Singing, Part III: Manual Garcia ii (1805–1906) and the Science of Singing in the Nineteenth Century." *Journal of Singing* 56, no. 4 (2000): 23–30.

Kietz, Hans. "Das räumliche Hören" [Spatial hearing]. *Acta Acustica United with Acustica* 3, no. 2 (1953): 73–86.

Kim-Cohen, Seth. *In the Blink of an Ear: Towards a Non-Cochlear Sonic Art*. New York: Continuum, 2009.

Kittler, Friedrich. *Gramophone, Film, Typewriter*. Translated with an introduction by Geoffrey Winthrup-Young and Michael Wutz. Stanford, CA: Stanford University Press, 1999.

Knapp, Rayond. "'Waitin' for the Light to Shine': Musicals and Disability." In *The Oxford Handbook of Music and Disability Studies*, edited by Blake Howe, Stephanie Jensen-Moulton, Neil Lerner, and Joseph Straus. Oxford University Press, forthcoming.

Knapp, Raymond, and Mitchell Morris. "Singing." In *The Oxford Handbook of the American Musical*, edited by Raymond Knapp, Mitchell Morris, and Stacy Ellen Wolf, 320–34. New York: Oxford University Press, 2011.

Knight, Wendy E. J., and Nikki S. Rickard. "Relaxing Music Prevents Stress-Induced Increases in Subjective Anxiety, Systolic Blood Pressure, and Heart Rate in Healthy Males and Females." *Journal of Music Therapy* 38, no. 4 (2001): 254–72.

Koenig, Carole. "Meredith Monk: Performer-Creator." TDR 20, no. 3 (1976): 51–66.

Koestenbaum, Wayne. *The Queen's Throat: Opera, Homosexuality, and the Mystery of Desire*. New York: Poseidon, 1993.

Kottman, Paul A. *A Politics of the Scene*. Stanford, CA: Stanford University Press, 2008.

Kramer, Lawrence. "'Longindyingcall': Of Music, Modernity, and the Sirens." In *Music of the Sirens*, edited by Linda Phyllis Austern and Inna Naroditskaya, 194–215. Bloomington: Indiana University Press, 2006.

———. "Music Criticism and the Postmodernist Turn: In Contrary Motion with Gary Tomlinson." *Current Musicology* 53 (1993): 25–35.

———. "The Musicology of the Future." *Repercussions* 1, no. 1 (1992): 5–18.

Krauss, Rosalind. "Video: The Aesthetics of Narcissism." *October* 1 (1976): 50–64.

Kreiman, Jody, and Diana Sidtis. *Foundations of Voice Studies: An Interdisciplinary Approach to Voice Production and Perception*. Malden, MA: Wiley-Blackwell, 2011.

Kuhn, Thomas S. *The Structure of Scientific Revolutions*. Chicago: University of Chicago Press, 1965.

Kreidl, A., and S. Gatscher. "Über die Lokalisation von Schallquellen" [On the localization of sound sources]. *Naturwissenschafte* 11, no. 18 (1923): 337–38.

Kumerdej, Mojca. "Sopranistko Juliano Snapper Pod Vodo, Sirenine Podvodne Arije" [Soprano Juliano Snapper's underwater sirensong]. *Delo*, June 28, 2008, 24–26. Translated by Luke Dunne and Juliana Snapper.

Kunimoto, Namiko. "Shiraga Kazuo: The Hero and Concrete Violence." *Art History* 36, no. 1 (2013): 154–79.

Kurihara, Kazutaka. "SpeechJammer." Accessed March 7, 2015. https://sites.google.com/site/qurihara/top-english/speechjammer.

Kurihara, Kazutaka, and Koji Tsukada. "SpeechJammer: A System Utilizing Artificial Speech Disturbance with Delayed Auditory Feedback." Accessed April 28, 2015. http://arxiv.org/abs/1202.6106.

LaBelle, Brandon. *Acoustic Territories: Sound Culture and Everyday Life*. New York: Continuum, 2010.

———. *Background Noise: Perspectives on Sound Art*. New York: Continuum International, 2006.

———. *Lexicon of the Mouth: Poetics and Politics of Voice and the Oral Imaginary*. New York: Bloomsbury, 2014.

Laozi. *Tao Te Ching: A New English Version*. Translated by Stephen Mitchell. New York: HarperCollins, 1988.

Laskowski, Larry. *Heinrich Schenker, an Annotated Index to His Analyses of Musical Works*. New York: Pendragon Press, 1978.

Le Guin, Elisabeth. *Boccherini's Body: An Essay in Carnal Musicology*. Berkeley: University of California Press, 2006.

Lehmann, Lilli, and Richard Aldrich. *How to Sing*. Rev. ed. New York: Macmillan Company, 1914.

Leopold, Aldo. *A Sand County Almanac: With Essays on Conservation.* New York: Oxford University Press, 1964.

Leppert, Richard. "Phonography and Operatic Fidelities (Regimes of Musical Listening, 1904-1929)." Paper presented at the University of Chicago Music Department's Colloquium Series, May 2, 2014.

———. "Reading the Sonoric Landscape." In *The Sound Studies Reader*, edited by Jonathan Sterne, 409–18. New York: Routledge, 2012.

Libin, Laurence, and Renato Meucci. "Organology." *The Grove Dictionary of Musical Instruments*, second edition, 3:753–55. New York: Oxford University Press, 2014.

Lindsay, R. Bruce. "Report to the National Science Foundation on Conference on Education in Acoustics." *Journal of the Acoustical Society of America* 36 (1964): 2241–43.

———. "The Story of Acoustics." *Journal of the Acoustical Society of America* 39, no. 4 (1966): 629–44.

Linklater, Kristin. *Freeing the Natural Voice: Imagery and Art in the Practice of Voice and Language.* Rev. and expanded ed. London: Drama, 2006.

Louboutin, S. R. Y. "Coordinated Group Musical Experience." *Google Patents*. US8521316 B2. Filed March 31, 2010, issued August 27, 2013. Accessed January 1, 2015, http://www.google.com/patents/US8521316.

Lortimer, Jan, dir., "Jam, Jelly and Juice," March 2013.

Lunenfeld, Peter. "The Maker's Discourse." Paper presented at the "Critical Mass: The Legacy of Hollis Frampton" conference, University of Chicago, IL, February 6, 2010.

Lyden, Jacki. "Remembering Aldo Leopold, Visionary Conservationist and Writer." *All Things Considered*. National Public Radio. March 10, 2013. Accessed March 1, 2015. http://www.npr.org/2013/03/10/173949498/remembering-aldo-leopold-visionary-conservationist-and-writer.

Mackenzie, Adrian. *Transductions: Bodies and Machines at Speed.* London: Continuum, 2002.

MacLarnon, Ann M., and Gwen P. Hewitt. "The Evolution of Human Speech: The Role of Advanced Breathing Control." *American Journal of Physical Anthropology* 109, no. 3 (1999): 341–63.

Malacart, Laura. "MUVE (Museum of Ventriloquial Objects): Reconfiguring Voice Agency in the Liminality of the Verbal and the Vocal." PhD diss., University College London, 2011.

Mann, Charles Riborg, and George Ransom Twiss. *Physics.* Chicago: Scott, Foresman and Company, 1910.

Marranca, Bonnie. "Performance and the Spiritual Life: Meredith Monk in Conversation with Bonnie Marranca." *Journal of Performance and Art* 31, no. 1 (2009): 16–36.

Massumi, Brian. *Parables for the Virtual: Movement, Affect, Sensation.* Durham, NC: Duke University Press, 2002.

———. *Semblance and Event: Activist Philosophy and the Occurrent Arts.* Cambridge, MA: MIT Press, 2011.

———. *What Animals Teach Us about Politics.* Durham, NC: Duke University Press, 2014.

Martin, Michel. "Pastor Jim Wallis Back to Being Political." *Tell Me More.* National

Public Radio. April 12, 2013. Accessed April 25, 2013. http://www.npr.org/2013/04/12/177032267/pastor-jim-wallis-back-to-being-political.

McCabe, Brian F. "Beethoven's Deafness." *Annals: Official Journal of the American Broncho-Esophagological Association* 67, no. 1 (1958): 192–220.

McCance, Dawne. "Crossings: An Interview with Kristin Linklater." *Mosaic* 44, no. 1 (2011): 1–45.

McCarthy, Elena. *International Regulation of Underwater Sound: Establishing Rules and Standards to Address Ocean Noise Pollution*. Boston: Kluwer Academic, 2004.

McClary, Susan. "Constructions of Subjectivity in Schubert's Music." In *Queering the Pitch: The New Gay and Lesbian Musicology*, edited by Philip Brett, Elizabeth Wood, and Gary Thomas, 205–34. New York: Routledge, 1994.

———. *Feminine Endings: Music, Gender, and Sexuality*. Minneapolis: University of Minnesota Press, 1991.

———. *Modal Subjectivities: Self-Fashioning in the Italian Madrigal*. Berkeley: University of California Press, 2004.

———. "Towards a Feminist Criticism of Music." *Canadian University Music Review* 10, no. 2 (1990): 19–18.

McClary, Susan, and Robert Walser. "Start Making Sense: Musicology Wrestles with Rock." In *On Record: Rock, Pop, and the Written Word*, edited by Simon Frith and Andrew Goodwin, 237–49. New York: Pantheon, 1990.

Medwin, Herman. *Sounds in the Sea: From Ocean Acoustics to Acoustical Oceanography*. Cambridge: Cambridge University Press, 2005.

Medwin, Herman, and Clarence S. Clay. *Fundamentals of Acoustical Oceanography*. Boston: Academic, 1998.

Meizel, Katherine. *Idolized: Music, Media, and Identity in American Idol*. Bloomington: Indiana University Press, 2011.

———. "A Powerful Voice: Investigating Vocality and Identity." In *A World of Voice: Voice and Speech across Culture, Voice and Speech Review* 7, no. 1 (2011): 267–74.

Menon, Vinod, and David J. Levitin. "The Rewards of Music Listening: Response and Physiological Connectivity of the Mesolimbic System." *NeuroImage* 28, no. 1 (2005): 175–84.

Mersenne, Marin. *Harmonie Universelle: The Books on Instruments*. Translated by Roger E. Chapman. The Hague, the Netherlands: Martinus Nijhoff, 1957.

Michelson, Annette, Richard Serra, and Clara Weyergraf. "The Films of Richard Serra: An Interview." *October* 10 (Autumn 1979): 69–104.

Middleton, Richard. *Studying Popular Music*. Buckingham, UK: Open University Press, 1990.

Miller, Dayton Clarence. *Anecdotal History of the Science of Sound to the Beginning of the 20th Century*. New York: Macmillan, 1935.

Mills, Mara. "Deaf Jam: From Inscription to Reproduction to Information." *Social Text* 28, no. 102 (2010): 35–58.

———. "Media and Prosthesis: The Vocoder, the Artificial Larynx, and the History of Signal Processing." *Qui Parle* 21, no. 1 (2012): 107–49.

Moi, Toril. *What Is a Woman? and Other Essays*. Oxford: Oxford University Press, 1999.

Mol, Annemarie. *The Body Multiple: Ontology in Medical Practice*. Durham, NC: Duke University Press, 2002.

Monson, Ingrid. "Hearing, Seeing, and Perceptual Agency." *Critical Inquiry* 34, no. 5 (2008): S36–58.

Montagu, Jeremy. *Origins and Development of Musical Instruments*. Lanham, MD: Scarecrow Press, 2007.

Moore, Robin. "The Decline of Improvisation in Western Art Music: An Interpretation of Change." *International Review of the Aesthetics and Sociology of Music* 23, no. 1 (1992): 61–84.

Moten, Fred. *In the Break: The Aesthetics of the Black Radical Tradition*. Minneapolis: University of Minnesota Press, 2003.

Müller, Viktor, and Ulman Lindenberger. "Cardiac and Respiratory Patterns Synchronize between Persons During Choir Singing." *PLoS ONE* 6, no. 9 (2011): e24893.

Muñoz, José Esteban. *Disidentifications: Queers of Color and the Performance of Politics*. Minneapolis: University of Minnesota Press, 1999.

"Nader: Obama Trying to 'Talk White' and 'Appeal to White Guilt.'" *Huffington Post*, July 7, 2008. Accessed April 19, 2015. http://www.huffingtonpost.com/2008/06/25/nader-obama-trying-to-tal_n_109085.html/.

Nancy, Jean-Luc. *Listening*. Translated by Charlotte Mandell. New York: Fordham University Press, 2007.

National Research Council Committee on a Conceptual Framework for New K–12 Science Education Standards. *A Framework for K–12 Science Education: Practices, Crosscutting Concepts, and Core Ideas*. Washington: National Academies Press, 2012.

Neumark, Norie, Ross Gibson, and Theo Van Leeuwen. *Voice: Vocal Aesthetics in Digital Arts and Media*. Cambridge, MA: MIT Press, 2010.

NickLane TV. "Speech Jammer." April 7, 2013. Accessed June 1, 2013. http://www.youtube.com/watch?v=Fzz7By8skLY.

"Notes & Queries," *Scientific American* 50, no. 14 (1884), 218.

O'Callaghan, Casey. *Sounds: A Philosophical Theory*. New York: Oxford University Press, 2007.

Ochoa Gautier, Ana María. *Aurality: Knowledge and Listening in Nineteenth-Century Colombia*. Durham, NC: Duke University Press, 2014.

———. "The Zoopolitics of the Voice in Colombia in the Nineteenth Century." The Eleventh Annual Robert Stevenson Lecture at University of California, Los Angeles, May 2, 2013.

O'Hagan, Sean. "Interview: Marina Abramović." *Observer*, October 2, 2010. Accessed April 1, 2010. http://www.theguardian.com/artanddesign/2010/oct/03/interview-marina-abramovic-performance-artist.

Oliveros, Pauline. *Deep Listening: A Composer's Sound Practice*. New York: iUniverse, 2005.

Omfgwhatisthisshit. "Speech Jamminn!!" October 27, 2012. Accessed June 1, 2013, http://www.youtube.com/watch?v=s_U_8x-Hiwc.

Ormandy, Eugene. Unpaginated foreword to Leo Leroy Beranek, *Music, Acoustics and Architecture*. New York: Wiley, 1962.

Oshima, Tetsuya. "'Dear Mr. Jackson Pollock': A Letter from Gutai." In *Under Each Other's Spell: Gutai and New York*, edited by Ming Tiampo, 13–16. Stony Brook, NY: Stony Brook Foundation, 2009.

Ouzounian, Gascia. "Embodied Sound: Aural Architectures and the Body." *Contemporary Music Review* 25, no. 1 (2006): 69–79.

Parker, Sybil P. *Acoustics Source Book*. New York: McGraw-Hill, 1988.

Patel, Rupal, and Anna Roden. "Intelligibility and Attitudes toward a Speech Synthesizer Vocoded Using Dysarthric Vocalizations." *Journal of Medical Speech-Language Pathology* 16, no. 4 (2008): 243–51.

Pennington, Stephen. "Transgender Passing Guides and the Vocal Performance of Gender and Sexuality." In *The Oxford Handbook of Queerness and Music*, edited by Fred Maus and Sheila Whiteley. New York: Oxford University Press, forthcoming.

Peraino, Judith. *Listening to the Sirens: Musical Technologies of Queer Identity from Homer to Hedwig*. Berkeley: University of California Press, 2006.

Pettman, Dominic. "Pavlov's Podcast: The Acousmatic Voice in the Age of MP3s." *differences* 22, nos. 2–3 (2011): 158.

Piekut, Benjamin. *Experimentalism Otherwise: The New York Avant-Garde and Its Limits*. Berkeley: University of California Press, 2011.

Pinch, Trevor, and Karin Bijsterveld, eds. *The Oxford Handbook of Sound Studies*. New York: Oxford University Press, 2012.

Pinch, Trevor, and Frank Trocco. *Analog Days: The Invention and Impact of the Moog Synthesizer*. Cambridge, MA: Harvard University Press, 2002.

Plato. *Symposium*. Translated by Alexander Nehamas and Paul Woodruff. Indianapolis, IN: Hackett, 1995.

Poizat, Michel. *The Angel's Cry: Beyond the Pleasure Principle in Opera*. Translated by Arthur Denner. Ithaca, NY: Cornell University Press, 1992.

Porzak, Simon, ed. "Liner Notes: The Margins of Song." Special issue. *Qui Parle* 21, no. 1 (2012).

Praetorius, Michael. In *Syntagma Musicum II, De Organographia*. Wolffenbüttel, Germany: Elias Holwein, 1619.

Price, Peter. *Resonance: Philosophy for Sonic Art*. New York: Atropos, 2011.

Pucci, Pietro. "The Song of the Sirens." *Arethusa* 12, no. 2 (1979): 121–32.

Pujol, Remy, Morello Lavigne-Rebillard, and Alain Uziel. "Development of the Human Cochlea." *Acta Oto-laryngologica* 111, no. s482 (1991): 7–13.

Rahaim, Matthew. *Musicking Bodies: Gesture and Voice in Hindustani Music*. Middletown, CT: Wesleyan University Press, 2012.

———. "Voice Cultures: Varieties of Ethical Power in Hindustani Music." Unpublished manuscript.

Rao, Singiresu S. *Mechanical Vibrations*. 5th ed. Upper Saddle River, NJ: Prentice Hall, 2011.

———. *Vibration of Continuous Systems*. Hoboken, NJ: John Wiley, 2007.

Redolfi, Michel. "Underwater Music Facts, Interview with Kevin McGuinness, Walt Disney Studio Home Entertainment, Sept 2008." Accessed April 2010. http://www.redolfi-music.com/eau/UnderwaterMusicFaqs.pdf.

Rehding, Alexander. "Of Sirens Old and New." In *Oxford Handbook of Mobile Music Studies*, edited by Sumanth Gopinath and Jason Stanyek, 2:77–108. Oxford: Oxford University Press, 2014.

"Researchers Build Directional 'SpeechJammer' Device." *Electronista*. March 1, 2012, accessed April 23, 2015, http://www.electronista.com/articles/12/03/01/system.said.to.work.without.physical.discomfort/.

Rhett and Link. "Speech Jammer Christmas Moments." December 19, 2012. Accessed June 1, 2013. http://www.youtube.com/watch?v=YNHRsOdZ3ig.

Riikonen, Tarina. "Shaken or Stirred—Virtual Reverberation Spaces and Transformative Gender Identities in Kaija Saariaho's *NoaNoa* (1992) for Flute and Electronics." *Organised Sound* 8, no. 1 (2003), 109–15.

Rituriel. "How to 'SpeechJam' with GarageBand." September 25, 2012. Accessed July 16, 2013. http://www.youtube.com/watch?v=Aosw7GI2tFc.

Rodgers, Tara. "Synthesizing Sound: Metaphor in Audio-Technical Discourse and Synthesis History." PhD diss., McGill University, 2010.

———. "Toward a Feminist Epistemology of Sound: Refiguring Waves in Audio-Technical Discourse." In *Philosophy after Irigaray*, edited by Mary Rawlinson, Danae Mcleod, and Sara McNamara. Albany: State University of New York Press, forthcoming.

Rosenberg, Harold. *The Tradition of the New*. New York: Da Capo, 1994.

———. "The American Action Painters." In Barbara A. Macadam, "Top Ten ARTnews Stories: 'Not a Picture but an Event.'" November 1, 2007. Accessed January 1, 2013, http://www.artnews.com/2007/11/01/top-ten-artnews-stories-not-a-picture-but-an-event/.

Ross, Charles D. "Blending History with Physics: Acoustic Refraction." *Chemistry and Physics Faculty Publications* Paper 26 (2000).

———. *Civil War Acoustic Shadows*. Shippensburg, PA: White Mane, 2001.

Russell, Bruce. "*A Priori* Justification and Knowledge." Stanford Encyclopedia of Philosophy Archive. Summer 2014. Accessed November 1, 2014. http://plato.stanford.edu/archives/sum2014/entries/apriori/.

Sakakeeny, Matt. "'Under the Bridge': An Orientation to Soundscapes in New Orleans." *Ethnomusicology* 54, no. 1 (2010): 1–27.

Saussure, Ferdinand de. *Course in General Linguistics*. Translated by Wade Baskin. New York: McGraw-Hill Book Company, 1966.

Schaeffer, Pierre. *Traité Des Objets Musicaux*. Paris: Éditions du Seuil, 1966.

Schaeffer, Pierre. *In Search of a Concrete Music*. Translated by Christine North, and John Dack. Berkeley: University of California Press, 2012.

Schlee, Susan. *The Edge of an Unfamiliar World: A History of Oceanography*. New York: Dutton, 1973.

Schlichter, Annette. "Un/Voicing the Self: Vocal Pedagogy and the Materialization of the Sounding Subject." *Postmodern Culture*, forthcoming.

Schwarz, David. "Listening Subjects: Semiotics, Psychoanalysis, and the Music of John Adams and Steve Reich." *Perspectives of New Music* 31, no. 2 (1993): 24–56.

Schwatz, Hillel. *Making Noise: From Babel to the Big Bang and Beyond.* New York: Zone, 2011.

Seeger, Anthony. *Why Suyá Sing: A Musical Anthropology of an Amazonian People.* Cambridge: Cambridge University Press, 1987.

Seiler, Emma. *The Voice in Singing.* Translated by William Henry Furness. Philadelphia: J. B. Lippincott, 1879.

Serra, Richard. *Boomerang.* 1974. Accessed April 1, 2012. http://www.ubu.com/film/serra_boomerang.html.

Serres, Michel. *The Five Senses: A Philosophy of Mingled Bodies.* Translated by Margaret Sankey and Peter Cowley. New York: Continuum, 2009.

Sewell, Stacey. "Listening Inside Out: Notes on an Embodied Analysis." *Performance Research* 15, no. 3 (2010): 60–65.

Shepherd, John, and Peter Wicke. *Music and Cultural Theory.* Malden, MA: Blackwell Publishers, 1997.

Shewell, Christina. *Voice Work: Art and Science in Changing Voices.* Hoboken, NJ: Wiley, 2009.

Small, Christopher. *Music, Society, Education.* Hanover, NH: University Press of New England, 1996.

———. *Musicking: The Meanings of Performing and Listening.* Middletown, CT: Wesleyan University Press, 1998.

Smith, Jacob. *Vocal Tracks: Performance and Sound Media.* Berkeley: University of California Press, 2008.

Smith, Mark M. *Listening to Nineteenth-Century America.* Chapel Hill: University of North Carolina Press, 2001.

Smithner, Nancy Putnam. "Meredith Monk: Four Decades by Design and by Invention." *TDR* 49, no. 2 (2005): 93–118.

Sontag, Susan. "Against Interpretation." In Susan Sontag, *Against Interpretation and Other Essays*, 3–14. New York: Farrar, Straus, and Giroux, 1966.

Stahl, Matt. "A Moment Like This: *American Idol* and Narratives of Meritocracy." In *Bad Music: The Music We Love to Hate*, edited by Chris Washburne and Maiken Derno, 212–32. New York: Routledge, 2004.

Stanley, Raymond M., and Bruce N. Walker. "Intelligibility of Bone-Conducted Speech at Different Locations Compared to Air-Conducted Speech." Paper presented at the annual meeting of the Human Factors and Ergonomics Society, San Antonio, TX, October 19–23, 2009.

Star, Susan Leigh, and James R. Griesemer. "Institutional Ecology, 'Translations' and Boundary Objects: Amateurs and Professionals in Berkeley's Museum of Vertebrate Zoology, 1907–39." *Social Studies of Science* 19, no. 3 (1989): 387–420.

Sterne, Jonathan. *The Audible Past: Cultural Origins of Sound Reproduction.* Durham, NC: Duke University Press, 2003.

———. "Listening Is One of the Most Enculturated Activities." *Weekend Edition.* NPR. September 14, 2002.

———. *MP3: The Meaning of a Format.* Durham, NC: Duke University Press, 2012.

———, ed. *The Sound Studies Reader.* New York: Routledge, 2012.

Stevens, Kenneth M., and William G. Hemenway. "Beethoven's Deafness." *Journal of the American Medical Association* 213, no. 3 (1970): 434–37.

Stockhausen, Julius. *A Method of Singing*. London: Novello, 1884.

Stoever-Ackerman, Jennifer. "Splicing the Sonic Color-Line: Tony Schwartz Remixes Postwar Nueva York." *Social Text* 28, no. 1 (2010): 59–85.

Sudnow, David, *Ways of the Hand: A Rewritten Account*. With an introduction by Hubert L. Dreyfus. Cambridge, MA: MIT Press, 2001.

Sujatha, Chandramohan. *Vibration and Acoustics: Measurement and Signal Analysis*. New Delhi: Tata McGraw Hill Education, 2010.

Sussman, Mark. "Fueled by Sentences: The Uncanny Art of Karl Ove Knausgaard." *Los Angeles Review of Books*. May 24, 2013. Accessed April 1, 2014. http://lareviewofbooks.org/review/fueled-by-sentences-the-uncanny-art-of-karl-ove-knausgaard.

Szendy, Peter. *Listen: A History of Our Ears*. Foreword by Jean-Luc Nancy, translated by Charlotte Mandell. New York: Fordham University Press, 2008.

Szwed, John F. *Space Is the Place: The Lives and Times of Sun Ra*. New York: Da Capo, 1998.

Taylor, Timothy D. *The Sounds of Capitalism*. Chicago: University of Chicago Press, 2012.

Thompson, Emily Ann. *The Soundscape of Modernity: Architectural Acoustics and the Culture of Listening in America, 1900–1933*. Cambridge, MA: MIT Press, 2002.

Thomson, Philip. *The Cambridge Companion to Brecht*, edited by Peter Thomason and Glendyr Sacks. Cambridge: Cambridge University Press, 2006.

Thurman, Judith. "Walking through Walls: Marina Abramović's Performance Art." *New Yorker*, March 8, 2010. Accessed April 1, 2014. http://www.newyorker.com/magazine/2010/03/08/walking-through-walls.

Tiainen, Milla. "Corporeal Voices, Sexual Differentiations: New Materialist Perspectives on Music, Singing and Subjectivity." *Thamyris/Intersecting* 18, no. 1 (2008): 147–68.

Tiampo, Ming. "Under Each Other's Spell: Gutai and New York." In *Under Each Other's Spell: Gutai and New York*, edited by Ming Tiampo, 3–12. Stony Brook, NY: Stony Brook Foundation, Inc., 2009.

Tippett, Krista. "Meredith Monk: Archaeologist of the Human Voice." *On Being*. American Public Media. February 16, 2012. Accessed April 1, 2014. http://www.onbeing.org/program/meredith-monko39s-voice/transcript/1401.

Tomlinson, Gary. "Evolutionary Studies in the Humanities: The Case of Music." *Critical Inquiry* 39, no. 4 (2013): 647–75.

———. "Musical Pasts and Postmodern Musicologies: A Response to Lawrence Kramer." *Current Musicology* 53 (1993): 18–24.

———. *The Singing of the New World: Indigenous Voice in the Era of European Contact*. Cambridge: Cambridge University Press, 2007.

———. "Tomlinson Responds." *Current Musicology* 53 (1993): 36–40.

Tongson, Karen. "Choral Vocality and Pop Fantasies of Collaboration." *Journal of Popular Music Studies* 23, no. 2 (2011): 229–34.

Tsukada, Koji. "SpeechJammer: Utilizing Artificial Speech Disturbance with Delayed Auditory Feedback." September 20, 2012. Accessed March 7, 2015. http://mobiquitous.com/speechjammer-e.html.

Treitler, Leo. "The Early History of Music Writing in the West." *Journal of the American Musicological Society* 35, no. 2 (1982): 237–79.

———. "The 'Unwritten' and 'Written' Transmission of Medieval Chant and the Start-Up of Musical Notation." *Journal of Musicology* 10, no. 2 (1992): 131–91.

Tresch, John, and Emily I. Dolan, "Toward a New Organology: Instruments of Music and Science." OSIRIS 28 (2013): 278–98.

Trower, Shelley. *Senses of Vibration: A History of the Pleasure and Pain of Sound*. New York: Continuum, 2012.

Truesdell, Clifford. *The Rational Mechanics of Flexible or Elastic Bodies, 1638–1788: Introduction to Leonhardi Euleri Opera Omnia, Vol. X et XI Seriei Secundae*, vol. 11, sec. 2. Zürich: Orell Füssli, 1960.

Tsai, Chen-Gia, Rong-Shan Chen, and Tzung-Shian Tsai. "The Arousing and Cathartic Effects of Popular Heartbreak Songs as Revealed in the Physiological Responses of Listeners." *Musicae Scientiae* (July 2014): 1–13.

Uffelen, Chris van. *Performance Architecture + Design*. Salenstein, Switzerland: Braun, 2010.

Urick, Robert J. *Principles of Underwater Sound*. New York: McGraw-Hill, 1983.

VanLoo, Babeth M., dir. *Inner Voice: Meredith Monk*. Hilversum, the Netherlands: House Foundation for the Arts, 2008. DVD.

Vennard, William. *Singing: The Mechanism and the Technic*. New York: Carl Fischer, 1968.

Vestinavesti. "Interview with Juliana Snapper." Accessed April 1, 2010. http://www.youtube.com/watch?v=gqDz6UCT5ws&feature=PlayList&p=06C40068B70D736D&playnext_from=PL&playnext=1&index=5.

Vickhoff, Björn, et al. "Music Determines Heart Rate Variability of Singers." *Frontiers in Psychology* 4 (2013): 1–16.

Wagner, Anne M. "Performance, Video, and the Rhetoric of Presence." *October* 91 (2000): 59–80.

Wald, Gayle. "Soul Vibrations: Black Music and Black Freedom in Sound and Space." *American Quarterly* 63, no. 3 (2011): 673–96.

Wallach, Hans, Edwin B. Newman, and Mark R. Rosenzweig. "The Precedence Effect in Sound Localization." *American Journal of Psychology* 62, no. 3 (1949): 315–36.

Wallis, Jim. *On God's Side: What Religion Forgets and Politics Hasn't Learned about Serving the Common Good*. Grand Rapids, MI: Brazo, 2013.

Weidman, Amanda J. *Singing the Classical, Voicing the Modern: The Postcolonial Politics of Music in South India*. Durham, NC: Duke University Press, 2006.

Westboro Baptist Church. "God Hates Fags." Accessed July 10, 2012. http://www.godhatesfags.com.

Wilson, H. A., and Charles S. Myers. "The Influence of Binaural Phase Differences in the Localization of Sound." *British Journal of Psychology* 2, no. 4 (1908): 363–85.

"Winners of the Ig Nobel Prize." Improbable Research. Accessed March 7, 2015. www.improbable.com/ig/winners.

Witherspoon, Herbert. *Singing, a Treatise for Teachers and Students*. New York: George Schirmer, 1925.

Wolfe, Joe, and Emery Schubert. "Did Non-Vocal Instrument Characteristics Influence Modern Singing?" *Musica Humana* 2, no. 2 (2010): 121–38.

Wolff, Janet. "Foreword: The Ideology of Autonomous Art." In *Music and Society: The Politics of Composition, Performance and Reception*, edited by Richard Leppert and Susan McClary, 1–12. Cambridge: Cambridge University Press, 1987.

Wong, Mandy-Suzanne. "Action, Composition—Morton Feldman and Physicality." 2009. Accessed April 1, 2013. http://eventalaesthetics.net/sonauto/Wong%20-%20Feldman%20Turfan.pdf.

Wyttenbach, Robert A., and Ronald R. Hoy. "Demonstration of the Precedence Effect in an Insect." *Journal of the Acoustical Society of America* 94, no. 2 (1993): 777–84.

INDEX

Abbate, Carolyn, 13, 34
Abramović, Marina, 33, 208n40
Acker, Kathy, 29
acoustic communities, 63–65
acoustic mediation, of voice, self, and others, 58–94
acoustic phenomena: figure of sound framework and, 169–78; fine tone quality and, 65–69; Lindsay's Wheel of Acoustics and, 165–69; listeners' choice and, 90–91; manipulation of, for prisoners, 189n8; music practices and, 23; in underwater singing, 33–34, 202n48
acoustic shadow, 195n63
action: action-based voice pedagogy and, 145–48, 225n22; intermaterial vibration and, 95–101, 139–40; sound-based voice lesson and, 140–45; vocal sounds as, 104–10
action painting, 101–4, 130–31, 216n37
Adorno, Theodor, 6, 53, 75, 78
Aeolian harp, 167–68
"Against Interpretation" (Sontag), 189n13
Alter Banhof Video Walk (Cardiff performance piece), 210n58
Althusser, Louis, 203n62
Amacher, Maryanne, 202n47
American Idol (television program), 225n28
Anderson, Laurie, 29, 32, 70
An die Nachgeborenen (To Those Born Later) (Brecht), 36

animal communication forms, Snapper's juxtaposition of opera with, 39–45, 200n29
a priori sound: defined, 6, 191n24; intermaterial vibrational practice and, 162–65; listening process and, 129–31
Aquaopera, 29
"Aquaopera #3/Los Angeles," 197n3
"AquaSonic" project, 198n11
Aristotle, 218n51
Arnheim, Rudolf, 209n49
art shows, Monk's work and, 72–80
Athey, Ron, 30–32, 36–38, 199n22
audience absorption coefficient, 64, 206n17
audience capacity, concert-hall acoustics and, 61–65, 205n7
audience participation: audience-musician interface and, 29, 31, 58–59, 239n107; in *Invisible Cities*, 83–90, 211n70
"audile techniques," 27
Audiothèque concept, 226n35
Augustine (Saint), 125–27, 221n79
Auner, Joseph, 14
aurality in music, 14, 190n14

Bacon, Francis, 163, 231n32
Barthes, Roland, 33, 128
bathroom, music performances in, 27–28, 197n2
Beauvoir, Simone de, 48–49, 51
Beethoven, Ludwig von, 149–50, 152–53

Bell, Gelsey, 29
Bennett, Jane, 16, 195n52
Beranek, Leo, 65–69, 206nn17–18, 206nn20–21, 207n25, 207n28
Berberian, Cathy, 29, 31–32, 39
Bergeron, Katherine, 226n35
Berio, Luciano, 39
Berkeley, George, 187n1
Bieletto, Natalia, 41, 44–45
biosensors, 177; use in *Body Music* of, 219n52
Blanchard, Elodie, 23–24, 104–10, 216n37
Blauert, Jens, 46, 175–78, 207n32, 235n73
Bleckmann, Theo, 81
body: as instrument, 110–20, 218n50, 232n40; intermaterial vibration and, 184–85; lived body concept, 47–52; musical experience and role of, 12–13; performativity and, 34; posture and body language, 222n90; pushing limits of, 29–31; sensing body and material world, 45–46; in Snapper's "hysterical" repertoire, 40–45, 198n5; vibrational theory of music and, 17–18; vibratory model of, 170–74
Body Music experiment, 111–20, 145, 219n56; experience and meaning in, 120–29; a priori sound and, 130–31
bone-conduction theory, 45–46, 201nn43–44
Bono, 134
Boomerang (Serra art piece), 95–101, 109, 124, 130, 212n2, 212n10, 212n13, 216n36
Boon, Marcus, 225n27
Born, Georgina, 189n8, 190n18
Boyle, Robert, 163
Braidotti, Rosi, 230n23
breathing: *Body Music* experiment and, 114–20, 219n56, 221n71; singing and, 219n53
Brecht, Berthold, 36
Breuer, Joseph, 200n31
Brown, Bill, 164
Brown, Oren, 225n25
Brunot, Ferdinand, 226n35
Burnett, Carol, 29
Butler, Judith, 48
Butler, Shane, 124, 158–59

Cage, John, 31, 33, 161
Calvino, Italo, 80, 82
Canguilhem, Georges, 217n45
Carnegie Hall (New York), 63–65
castrato voice, 13
Cavarero, Adriana, 53, 99, 101
"cello-and-bow thinking," 14
Cerrone, Christopher, 23, 58–59, 80–90, 92
"Challenging Mud" (Shiraga), 216n32
charity, vibrational theory of music and, 21
Chase, Nick, 216n37
Chion, Michel, 149, 151, 227n42
choreography, *Body Music* experiment and, 115–20
Chou, Pai, 111
Christianity, in Snapper's performances, 34–38
cognitive science, vibrational theory of music and, 191n32
collective rehearsal process, 105–10
Collins, Billy, 4–5, 189n13
A Complete Treatise on the Art of Singing (Garcia), 136
composition: experience and meaning in, 120–29; by singer-composers, 29
concert-hall acoustics: fine tone quality and, 65–69, 206n19; future challenges in, 212n76; history of, 60–65, 205n4, 205n7; in *Invisible Cities*, 85–90; listeners' choice and, 90–91; Monk's work and, 74–80; psychoacoustics and, 68–69, 207n32
Confessions (Augustine), 125–26
Connor, Steven, 34, 44, 150, 196n72
conservatory acoustics, music pedagogy and, 223n11
Copenhafer, David Tyson IV, 52, 203n63
copying, in music, 225n27
Corbin, Alain, 14
Cristchurch Town Hall for Performing Arts, 205n8
Cronan, Paula, 32
Cruz, Jon, 6
Cuddy, Amy, 222n90
cultural construction: audile techniques and, 27–29; listening and, 100–101, 222n92;

lived body and, 48–52; music and, 233n.46; performativity and, 34; vibrational theory of music and, 10–16, 18–19
Cusick, Suzanne, 34, 189n8

Dart, Thurston, 63
Davies, James Q., 14, 218n48
Davies, Peter Maxwell, 39–40
Davis, Vaginal Creme, 32
deaf culture, media and, 14
Debussy, Claude, 53
DeNora, Tia, 112, 159–60, 171, 218n50, 219n54, 222n91
Derrida, Jacques, 97, 121–24, 127–28
Dialogues Concerning Two New Sciences (Galileo), 163
Dillon, Emma, 125–27, 158–59, 193n38, 221n79
direct sound, Amacher's theory of, 202n47
disability studies, music and, 13–16
disidentification, Muñoz's concept of, 8–9
Dolan, Emily, 13, 164, 232n41
Dolar, Mladen, 52, 104–10, 145, 215n19
drastic vs. gnostic dilemma, voice studies and, 13, 193n38
Duncan, Michelle, 48, 50
duration, vibrational theory of music and, 17
Dyson, Frances, 135–36, 139–40

ear anatomy, sound transmission and transduction and, 170–72
Earth Sound Earth Signal, 234n59
Eastman, Julius, 40
echo, vocal ontology and, 95–101
Edison, Thomas, 150–53
8 Songs for a Mad King (Davies), 40
Einstein, Albert, 154
Eisenstadt Castle, 205n4
Eisler, Hans, 36
Electronic Arts Intermix (EAI), 96–101
energy states: figure of sound framework and, 169–78; intermaterial vibrational practice and, 165–69; music and, 194n50
Enrrationes in Psalmos (Augustine), 126
Erlmann, Veit, 14

Erwartung (Schoenberg), 39, 200n31
Esposito, Christina M., 220n63
ethics, sound studies and, 6, 191n26
"Evolutionary Studies in the Humanities: the Case of Music" (Tomlinson), 160
experience, limitations of, 120–29

Feldman, Martha, 13–14
Felman, Shoshana, 48
feminist scholarship: materiality in, 48–52; music and, 14–16, 198n13, 230n23; Snapper's performances and, 40–45; sound and, 199n1, 202n53
Fibonacci sequence, in music, 4
fidelity: of sound, 129; vibrational theory of music and, 21
figure of sound framework, 2; acoustics and, 90–94; *Body Music* experiment and, 129; concert-hall acoustics and, 60–65; defined, 187n4; fine tone quality and, 65–69; intermaterial vibrational practice and, 161–65; material practice and, 152–53; perspectives on vibration, transmission, and transduction and, 169–78; vibrational theory of music and, 17–22, 156–85
fine tone quality, 65–69
Fink, Robert, 19, 63, 68, 207n33
Finley, Karen, 32
Fitzmaurice, Catherine, 146
Five Fathoms Deep My Father Lies (Snapper), 29–31, 34–38, 197n3
Five Fathoms Deep Opera Project, 39–45, 197n3
Ford, Phil, 158–59
form-oriented analysis, in music studies, 13, 193n36
Forsyth, Michal, 205n4
Foucault, Michel, 109–10
Foundations of Voice Studies: An Interdisciplinary Approach to Voice Production and Perception (Kreiman and Sidtis), 12
Freud, Sigmund, 200n31

Galás, Diamanda, 29, 32
Galileo Galilei, 163, 169, 231n32

Garcia, Manuel, Jr., 136
Geertz, Clifford, 1, 4, 120, 187n2
Geissinger, Katie, 71
gender: lived body and, 47–52; in Snapper's "hysterical" repertoire, 40–45, 54–57; in voices, 218n47; voice studies and, 195n59; water medium and, 202n55
genre parameters, timbre and, 134–36
Germain, Sophie, 169
Gesamtkunstwerk, *Body Music* experiment, and, 118–20
Gimenez, E. Martin, 80–90, 92
glossolalia, in underground singing, 36–38
"God Hates Fags" campaign, 199n19
Goehr, Lydia, 133–36
Goodman, Steve, 53, 57
Gordon, Bonnie, 163–65
"The Grain of the Voice" (Barthes), 33
Great Treatise on Supreme Sound, 195n52
Griesemer, James, 164
Griffin, Sean, 32
Grosser Musikvereinsaal (Vienna), 63
Gutai (Concrete) Art Association, 103–10, 215n31

Hagen, Nina, 29
Hall, Stuart, 8–11
Hamilton, Ann, 72–74
håndarbeid (work of the hand), 15
Hanover Square Rooms, 61
harmony, in music studies, 13, 193n36
Harrison, Lou, 33
Harvey House Restaurant, 83–85
headphones, in *Invisible Cities* performances, 80–90, 211n68
heart rate, singing and, 219n53
Heidegger, Martin, 50, 182
Helmreich, Stefan, 194n51
Henao, Luis Fernando, 111
Henriques, Julian, 158
Hewitt, Gwen P., 221n71
Hirsch, Shelly, 29
Hirschkind, Charles, 105–52
Holsinger, Bruce, 126–27

Holt, Nancy, 95–101, 213n10, 216n36
Homer, 23, 51–57, 203n62
Horkheimer, Max, 53, 75, 78
House Foundation (Monk), 71
Hughes, Robert, 102–3
human entity, vibrational theory of music and, 184–85, 239n105
Human Microphone Project, 29
Hurricane Katrine, Snapper's underground performance project and, 34–36, 197n3
"hysterical" repertoire of Snapper, 40–45
hystericism, Snapper's concept of, 40, 200n31, 200n34

I Am Sitting in a Room (Lucier), 212n2
Ihde, Don, 211n72
Ilano, Roxanne "Rox," 197n2
Iliad (Homer), 203n61
The Industry opera company, 80
inevitability narrative, challenges to, 48–52
Infanti, Andrew, 32
infrasound dimensions, 171–78
inner choreography, voice as action and, 139–40
Inner Voice (documentary), 71
Institute for Research and Coordination in Acoustics/Music (IRCAM), 207n33
instruments and instrumentalization: audience-musician interface and, 29, 31, 58–59; body as, 110–20, 218n50, 232n4; in clothing, 104–10; concert-hall acoustics and, 63–65, 68–69; figure of sound framework and, 9, 169–78, 235n66; fine tone quality and, 65–67; historical artifacts of, 233n52; intermaterial vibration and, 165–70; in Monk's performances, 75; organological analysis and, 163–65, 233n44; Snapper's breakdown of, 31–34; work concept and, 135–36, 139–40
interdisciplinary performance, Monk's work as, 70–80
interiority, voice studies and, 195n57
intermaterial vibrational practice, 161–65, 234n59; energy and, 165–69; organological analysis of, 171–72

Invisible Cities (Cerrone), 23, 58–60, 80–94, 211nn70–71
invisible opera, *Invisible Cities* as, 85–90
Iverson, Paul, 220n63

Jaffe, Christopher, 64, 66, 205n8, 207n24
Jakobson, Roman, 121–23, 128, 131
Jankélévitch, Valdimir, 13
Jones, Amelia, 37, 103
Jousse, Marcel, 150–52
The Judas Cradle, 31, 36–38

Kafka, Franz, 52–57
Kahn, Douglas, 167–68, 234n59, 237n84
Kane, Brian, 178
Kaprow, Allan, 216n33
Kelman, Ari, 206n15
Knapp, Raymond, 110, 231n29
Knausgård, Karl Ove, 154
knowledge domains, in music, 4
Koestenbaum, Wayne, 48
Kramer, Lawrence, 53
Krauss, Rosalind, 96–97
Kreiman, Jody, 12
Kuhn, Thomas, 50–51
Kurihara, Kazutaka, 98–99

La Barbara, Joan, 29, 31–32
L.A. Dance Project, 80
Laennec, R. T. H., 6
"Land Ethic" (Leopold), 21–22
Laozi, 154
laryngoscope, 136
La Scala Opera House, 62
Lavender Mist (Pollock), 102
La Voix Humaine (Poulenc), 39
Le Guin, Elisabeth, 14
Lehrich, Christopher, 158
Leopold, Aldo, 21–22, 196n69
Leppert, Richard, 213n12
Lewis, George, 29
Linderberger, Ulmann, 219n53
Lindsay, R. Bruce, 165–69

Lindsay's Wheel of Acoustics, 165–69
linguistics: *Body Music* experiment and, 121–29; concert-hall acoustics and, 62–65; limits of the symbolic and, 159; underwater singing and, 40–45
Linklater, Kristin, 143–45, 226n34
Lippman, Rebecca, 16–18, 129
listening: acoustics and, 90–91; enculturation of, 100–101, 222n92; as intermaterial vibrational practice, 3; multisensorial listening, 148–52; multisensory perspective on, 52–57; a priori sound and, 129–31; theories of, 5–8; as thick event, 181–83
lived body, vocal tradition and, 47–52
Live in hd, 81
locatedness/localization, figure of sound framework and, 175–78, 235n73
Lockwood, Annea, 31
London Royal Opera, 62
Lucier, Alvin, 212n2
Lunenfeld, Peter, 14

MacKenzie, Adrien, 194n51
MacLarnon, Ann M., 221n71
madness, in Snapper's repertoire, 40
magical hermeneutics, 158
Malacart, Laura, 97
Mann, Elena, 32
materiality: dependency of music on, 27–28, 55–57; figure of sound and, 152–53; intermaterial vibrational practice, 161–65; lived body and, 48–52; music and listening and, 3, 14–16; sensing body and, 45–46
Matheson, Susan, 197n3
McClary, Susan, 15, 49–50, 57
meaning: limitations of, 120–29; limits of the symbolic and, 157–60
measurement, in music, 4
medieval music, 125, 158
The Medium (Davies), 39
Mehta, Zubin, 63
Meredith Monk & Vocal Ensemble, 58, 71
metaphorical approaches, in vocal pedagogy, 136–38

metaphysics of music, 15–16
Metropolitan Opera, 81, 209n54
microphones: in opera performances, 81, 209n54; piezo technology and, 177
Midgette, Anne, 69–70
Miller, George Bures, 210n58
Mills, Mara, 14
modernism, in music studies, 14
Moi, Toril, 49, 51
Mol, Annemarie, 20–21
"Monitor Unit for Solid Vibration" (Toshiya performance piece), 237n83
Monk, Meredith, 23, 29, 31–32, 58, 69–80, 90–92, 182; *Songs of Ascension* by, 72–80
morality, sound studies and, 6, 191n26
Moritz, Jonathan, 216n37
Morris, Mitchell, 110
Moten, Fred, 196n71
Müller, Victor, 219n53
multisensorial phenomenon: case studies in, 24–25; learning and, 148–52; listening and, 52–57; Monk's work as, 70–80; music as, 3–8; a priori sound and, 129–31
Munich Staatsoper, 62
Muñoz, José Esteban, 8–11
music: definitions of, 155; as multisensorial phenomenon, 3–8; naming process in, 24; naturalized cornerstones of, 8–16, 132–33, 191n31
musicking: action-based voice lesson and, 145–48; multisensorial listening and, 151–52; redefinition of music as, 5; as "Silent Practice," 171
musicology: challenges to traditional theories in, 49–52; new theories in, 15–16
music psychology, vibrational theory of music and, 191n32

Nader, Ralph, 195n61
naming process, in music studies, 24
Nancy, Jean-Luc, 183–84, 238n104
Naples San Carlo Opera House, 62
National Endowment for the Arts, 70
National Research Council, 169

naturalized parameters of music, 8–16, 63–64, 191n31; acoustics and, 90–94; fine tone quality and, 66–69; organological investigation and, 175–78; signifying sound and, 100–101; sound-based voice lesson and, 141–45
Neuhaus, Max, 33
Newton, Isaac, 169
Noisy Clothes musical experiment, 104–11, 129, 145, 216n37, 217n41
Noisy Clothes Suite, 216n37
nondiastematic notation, 125
Norderval, Kristin, 29, 32
normative listening practices, Snapper's challenge to, 48–57
notation, composition and, 120–29

objects, intermaterial vibration and, 164–65
"Occupy Wall Street" protest movement, 29
Ochoa Gautier, Ana Maria, 198n6
Odyssey (Homer), 23, 51–57, 203n62
Oleson, Jeanine, 32
Oliveros, Pauline, 29, 31, 70
"On Popular Music" (Adorno), 6
opera and operatic technique: body music and, 110–20, 220n57; in *Invisible Cities*, 80–90; Snapper's commitment to, 39–45; symbolic in relation to, 157–60
orchestral timbre, 13
organization of sound, music and, 5
organological inquiry: in intermaterial vibrational practice, 161–65, 175–78, 231n32; signifying sound and, 184–85

Pappenheim, Bertha, 200n31
Pappenheim, Marie, 200n31
paradigm shift, in music theory, 50–52
parallelepipedic architecture, concert-hall acoustics, and, 62–65
Paris Garnier Opera House, 62
Pavarotti, Luciano, 134
Pennington, Stephen, 218n47
Peraino, Judith, 203n62
Perey, Erin "Rin," 197n2

performance studies: acoustic mediation and, 58–94; *Body Music* experiment and, 112–20; concert-hall acoustics and, 60–65; external frames in, 217nn42–43; live streaming of opera performances and, 209n54; music and, 8–9; Pollock's action paintings and, 101–4; preservation of performance art, 71, 208n40; supermusicality and, 125–26; vibrational theory of music and, 23–24; vocal pedagogical frameworks and, 136–40; vocal sounds as action and, 104–10; work concepts and, 133–36

performativity: culture and, 34; in Monk's work, 70–80; multisensorial listening and, 151–52

Philadelphia Academy of Music, 62

philosophy of sound, lived body and, 50–52

phōnē (the voice), 120–22, 124–25

phonemes, 121–23, 220n63

phonometrograph, 14

Piekut, Benjamin, 178

piezo ceramic sensors, 177–78, 235n66

Pinch, Trevor, 164

pitch coherence: multisensorial listening and, 148–52; in music studies, 13, 193n36, 224n18; sound-based voice lesson and, 141–45; vocal pedagogy and, 138–40, 225n26

Plantamura, Carol, 32

Plato, 238n101

poetry: music as, 183–84, 238n101, 238n104; teaching and knowledge production in, 4–5

Poizat, Michel, 157–59, 229n6, 230nn19–20

Pollock, Jackson, 24, 101–5, 109, 130–31, 146, 215n31, 216n33

Pommier, Gérard, 230n20

popular culture, vocal sounds and, 225n28

posture, *Body Music* experiment, and, 114–20, 222n90

Poulenc, Francis, 39

power structures: sound-based voice lesson and, 141–45; sound process and, 109–10

practice, music as, 20–22

Prelude to the Holy Presence of Joan d'Arc (Eastman), 40

Principles of Human Knowledge (Berkeley), 187n1

prisoners, manipulation of acoustical environment for, 189n8

psychoacoustics, 46; concert hall design and, 68–69, 207n32

psychoanalytic theory, voice within, 159–61, 230n23

Pucci, Pietro, 203n61

Puckette, Miller, 32

pure sound, defined, 64, 206n15

Pythagoras, 169

Quran, listening stances in, 151–52

race studies, music and, 8–9, 14–16, 192n33, 195n57, 196n71, 202n55, 204n1, 228n54

Radio City Music Hall, 209n49

Rao, Singiresu S., 168–69, 237n92

Rayleigh (Lord), 169

Redolfi, Michel, 33

reggae sound systems, 158

Rehding, Alexander, 204n71

relationality: Born's musicology and, 190n18; thick event and, 181–83; vibrational theory of music and, 179–81

Renaissance philosophy, *musica*, and, 163

reverberation times: concert-hall acoustics, 63; fine tone quality and, 65–69, 206n17, 207n33

rhythm, in music studies, 13, 193n36

Rodgers, Tara, 202n53

Romantic work concept: vocal pedagogy and, 140; voice in service of, 133–36

Rosenberg, Harold, 103–5

Ross, Charles, 195n63

Rousseau, Jean-Jacques, 127–28

Royal Albert Hall (London), 61

Russolo, Luigi, 183

Ryle, Gilbert, 187n2

Satie, Eric, 14

Saussure, Ferdinand de, 121–24, 128

Schaeffer, Pierre, 227n42
Schenker, Heinrich, 6
Schlichter, Annette, 143–45, 226nn31–32
Schoenberg, Arnold, 39, 200n31
Schubert, Emery, 224n18
Schwarz, David, 159
Scientific American, 187n1
scoring, for *Body Music* experiment, 117–20
sensing body: material world and, 45–46; sound beyond aurality and, 55–57
Serra, Richard, 95–101, 213n8
Serres, Michel, 44
sexuality, lived body and, 47–52
Sharon, Yuval, 80–90, 92
Shepherd, John, 231n29, 238n100
Shiraga Kazuo, 216n32
shoebox concert hall model, acoustics and, 62–65, 68
Sidtis, Diana, 12
signifying sound, 100–101; *Body Music* experiment and, 121–29; limits of, 157–60; organological inquiry and, 184–85
signs, power and effect of, 8–9
silence: feminist scholarship on, 49–52; intermaterial vibrational practice and, 161; music and absence of, 52–57
silent disco events, 80–81
Sinatra, Frank, 134
singer-composers, 29
singing: as action, 95–101; alleged microphone use in, 209n54; *Body Music* experiment and, 112–20; dominant concepts of, 11–12; as intermaterial vibrational practice, 3; sound-based voice lesson and, 140–45; as thick event, 181–83; vibrational theory of music and, 23–24; vocal anatomy and, 110; vocal pedagogy and, 136–37; words and melody in, 127–29
Singing, the Mechanism and the Technic (Vennard), 136–37
single-character tradition, Snapper's performances and, 39–45
"Sirènes" (Debussy), 53
sirens, Odysseus and, 52–57, 203n62
Siskin, Clifford, 63

site specificity, Monk's work and, 74–80
Skype, 98–99, 212n13
slavery, music and, 6
Small, Christopher, 5, 15–16, 33–34, 151, 190n15
Smith, Mark, 6
Smith, Wadada Leo, 239n106
Snapper, Juliana, 23; challenges to inevitability narrative by, 47–52; on flood and rapture, 34–36; operatic technique and, 39–45, 182, 198n5; underwater performances by, 197n3; vibrational perspectives on, 27–57, 161; on vocal breakdown, 37–38
Sniffin, Allison, 71
social dynamics of music, 15–16, 233n46
Songs of Ascension (Monk), 23, 58–60, 72–80, 90–94
sonic experience, 2, 93–94, 108–10; sound-based voice lesson and, 141–45
sonic logos, Henriques' concept of, 158
sonic saddle, 231n29, 238n100
Sontag, Susan, 189n13
Soper, Kate, 29
sound: acoustic mediation of, 58–94; as action, 104–10; concert-hall acoustics and, 60–65, 205n15; direct sound theory, 202n47; in *Invisible Cities*, 85–90; in performance art, 23–24; research sources for study of, 188n7; transmission of, 200n38; vibrational theory of music and, 5–16, 156–85; voice pedagogy based on, 147–48
sound-based voice lesson, 140–45
spatial-relational dimensions, 213n10; acoustic mediation and, 60–65; Monk's work and, 73–80; music practices and, 23; organological analysis of sound and, 172–78
SpeechJammer device, 98–101, 109, 124, 214n16
Sprinkle, Annie, 32
stable work concept, vocal pedagogy and, 136–40
Stanley, Raymond M., 201n44
Star, Susan Leigh, 164
Sterne, Jonathan, 6, 27, 51, 139–40
stethoscope, invention of, 6

268 · INDEX

Stoever-Ackerman, Jennifer, 222n92
Strozzi, Barbara, 29
subjectivity, voice studies and, 195n57
Sudnow, David, 110
Sujatha, Chandramohan, 170–71, 176, 236n81, 237n85
Sun Ra, 162
supermusicality, 125–26
Swed, Mark, 210n58
symbolic, limits of, 157–60
Symphony Hall (Boston), 63
Szendy, Peter, 5

Taiainen, Milla, 230n23
Tavares da Rocha, João, 43
Taylor, Cecil, 31
technology, voice studies and, 13–14
The Tempest (Shakespeare), 36
textual authority, voice as, 199n13
thick events: *Body Music* experiment and, 119–20; in ethnography, 1–2, 187n2; limits of symbolic and, 157–60; music and, 5–8, 164–65; relationality and, 181–83
Thompson, Emily Ann, 75, 206n15, 206n19, 209n49
Thoreau, Henry David, 167
timbre: musical genres and, 134–36; vocal pedagogy and, 225n26
Tomlinson, Gary, 134, 160
Toshiya, Tsunoda, 237n83
transduction of sound: figure of sound framework and, 169–78; intermaterial vibration and, 164–67; music as node of, 16–22, 163
transferable energy, in music, 16
transgender voices, 218n47
transmission of sound: figure of sound framework and, 169–78; intermaterial vibration and, 164–67, 234n62; in *Invisible Cities*, 85–90; music as node of, 16–22
Treitler, Leo, 125, 133–36
Triana, Alba Fernanda, 24, 111, 121, 125
Trocco, Frank, 164
Trower, Shelley, 164
Tsukada, Koji, 98–99

ultrasound dimensions, 171–78
underwater singing, 11, 23, 194n51; material dependency of music and, 27–28, 55–57; operatic technique and, 39–45; participatory experiments in, 41–45; risks of, 200n29; vocal breakdown and, 37–38
Union Station (Los Angeles), 82–90
upside-down singing, 29–31, 36–38

Velcro, 217n41
Veludo, Hugo, 43
Vennard, William, 136–37
ventriloquism, 196n72
vibrational theory of music, 3, 7–16; figure of sound framework and, 169–78; intermaterial practice and, 161–65; multisensorial listening and, 148–52; music as vibrational practice, 154–85; musicology and, 16; naturalized parameters and, 8–16, 191n31; nodes of transmission and transduction and, 16–22; relationality and, 179–81; sensing sound and, 56–57; underwater singing and, 45–46
vibratory impact, standards for, 235n68
Vickhoff, Björn, 219n53
video art, 95–101
Vienna Staatsoper, 62
Visage (Berio/Berberian), 39
visual art, music and, 101–4, 213n12
vocabulary, body movements as, 220n58
vocal cords: anatomy of, 110–11, 224n17; *Body Music* experiment and, 116–20, 218n49
Vocal Ensemble. *See* Meredith Monk & Vocal Ensemble
vocal paradigms: contextual influences on, 31–34; a priori sound and, 129–31; singing and, 12; traditional *vs.* multisensory perspectives, 52–57
vocal paralysis and ontology, 95–101
VocalSynth, 32
vocal training, 14–15, 194n46; action-based voice pedagogy, 145–48, 225n25; Monk's work and, 70–80; multisensorial listening and, 148–52; of Snapper, 39–45, 198n5;

INDEX · 269

vocal training (*continued*)
   sound-based voice lesson, 140–45; symbolic in relation to, 157–60; work concepts' impact on, 136–40
voice: acoustic mediation of, 58–94; anatomy of, 110–20, 218n51; common conceptions about, 27–29; intermaterial vibration and, 161–65; as knowledge object, 2–3; lived body and breakdown of, 47–52; in musical events, 13; a priori sound and, 129–31; pushing limits of, 29–31; research sources for study of, 188n7, 198n6, 199n13; Snapper's breakdown of, 36–38; teaching approaches for, 14–15, 194n46; vocal sounds as action, 104–10; in vocal studio, 194n46; work concept and, 133–36
voice work, 143–45
VOX: Showcasing American Composers, 80

Wagner, Anne, 96–97
Walker, Bruce N., 201n44
Wallis, James, 20
Walt Disney Concert Hall, 58, 63, 78–79
Warner, William, 63
Watson, Thomas, 167
Weidman, Amanda, 135–36
Weiss, Sidney, 63
Westboro Baptist Church, 199n19
Western music parameters, 27–29; acoustics and, 91–93; concert-hall acoustics and, 60–65; notation and, 125; sound-based voice lesson and, 141–45; underground singing and, 27–29; vocality and, 132–33; vocal pedagogy and, 138–39; voice in service of, 133–36
Wicke, Peter, 231n29, 238n100
Wolfe, Joe, 224n18
work concepts: vocal pedagogy and, 136–40; voice in service of, 133–36

X, Amy, 29

York Buildings, 61
Yoshihara, Jirō, 103–4
You Who Will Emerge from the Flood, 29, 43, 46, 197n3

Z, Pamela, 29, 32
Zhang Heng, 169

www.ingramcontent.com/pod-product-compliance
Lightning Source LLC
Chambersburg PA
CBHW050211240426
43671CB00013B/2288